First Steps

to

Mathematics

for Infants

By

Cheryl .M. Greenidge

First edition 2017

Cover design by Cheryl .M. Greenidge
Edited by Jerome Davis
Published by Cheryl .M. Greenidge

ISBN-13: 978-1545501269

ISBN-10: 1545501262

ABOUT THE AUTHOR

Cheryl Greenidge attended St. Martin's Girl's School, the Princess Margaret Secondary School and the Barbados Community College. In 1988, she started her career as a primary school teacher. In 1997, she enrolled at the Erdiston Teacher's Training College where she completed her Diploma in Education. In 2005, Cheryl was made Early Childhood Coordinator at the St. Martin's - Mangrove Primary School. Cheryl's years of experience in the infants' department have greatly assisted her in compiling the material for this book. Cheryl is the author of 'Word Building for Infants', 'A Spelling and Reading Aid for Beginners' , 'Grammar Made Easy for Infants' – Books 1,2,3 & 4 and 'Number Bonds for Infants'.

CONTENTS

Part one

Part two

Part three

Preface

First Steps to Mathematics for Infants is a three-part book which introduces the young learner to Mathematics. The concepts are presented in an easy and interesting manner with a variety of activities for reinforcement.

The first part introduces concepts like left, right and middle, same and different, bigger and smaller, patterns, light and heavy, tall, short and long, full and empty and holds more and holds less.

The second part focuses on numbers zero to ten. The activities include tracing and writing the numbers, drawing objects for the numbers, circling and completing sets, and counting and matching to the number.

The third part presents concepts like shapes, one to one correspondence, same, equals, more than and less than, bigger and smaller number, missing numbers in series 0-10, number names zero to ten, ordinals, whole and fractions, pictographs, classifying, and an introduction to adding and taking away.

Although designed for the 4-6 age group, it may be helpful to older children who have not mastered the concepts.

Part 1

✶ Circle and then colour the pictures that are on the left.

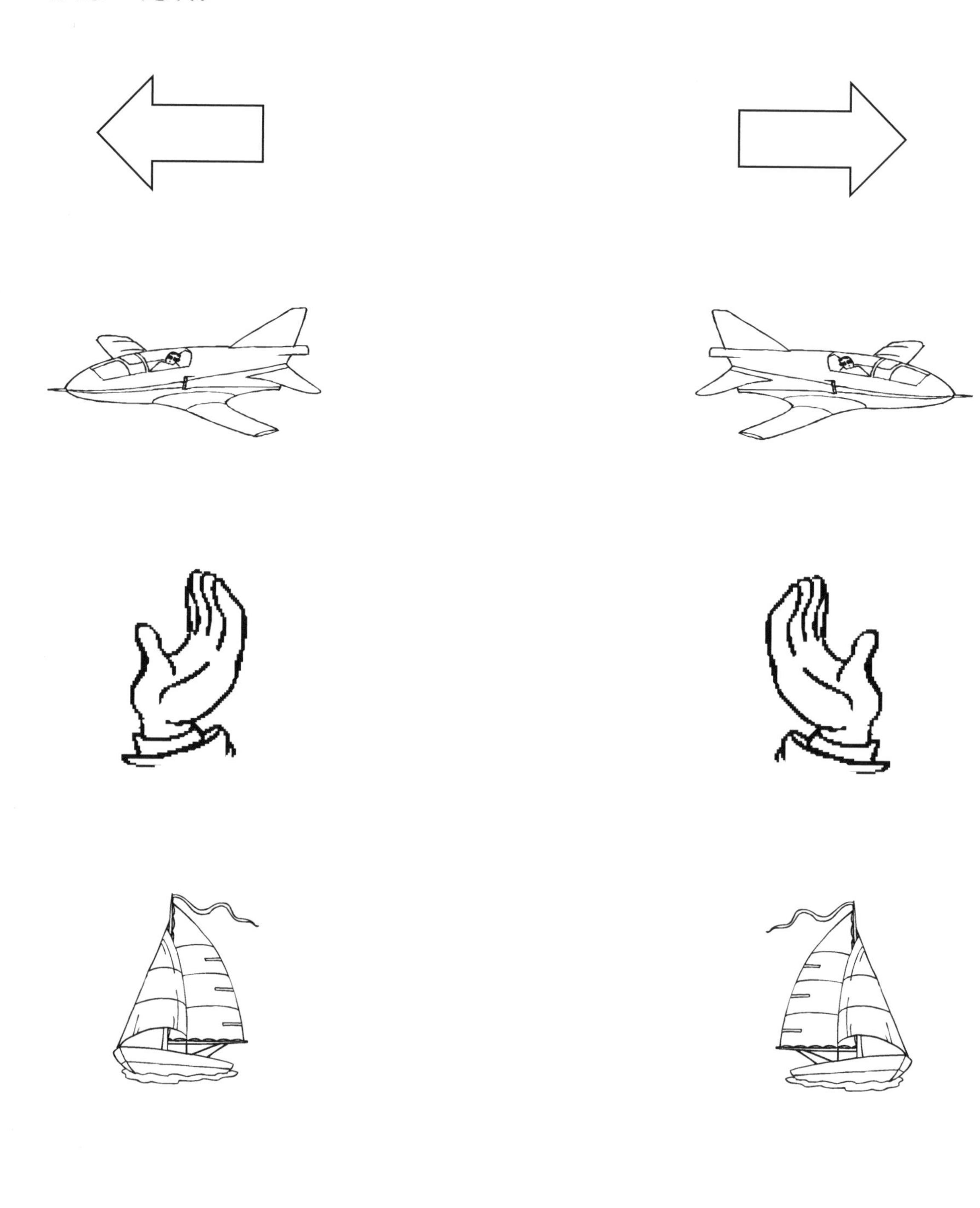

✵Circle and then colour the pictures that are on the right.

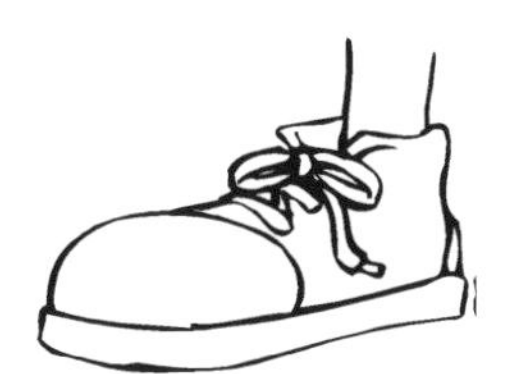 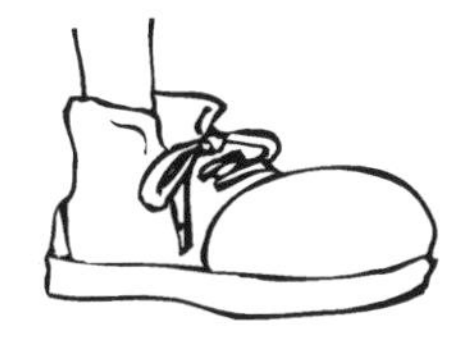

✲ Circle and then colour the pictures that are facing the left.

✲ Circle and then colour the pictures that are facing the right.

 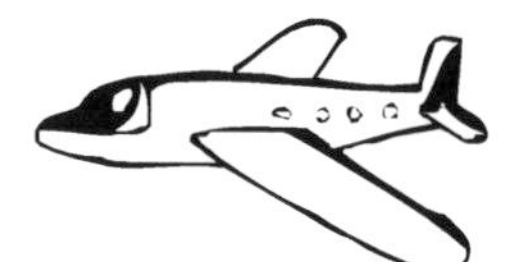  

✱In each box, circle and then colour the picture that is in the middle.

✱In the box draw: a ball in the middle

a cup on the left

a cat on the right

✱Match the pictures that are the same.

✵In each line, circle the two pictures that are the same.

*In each box, circle and then colour the picture that is different.

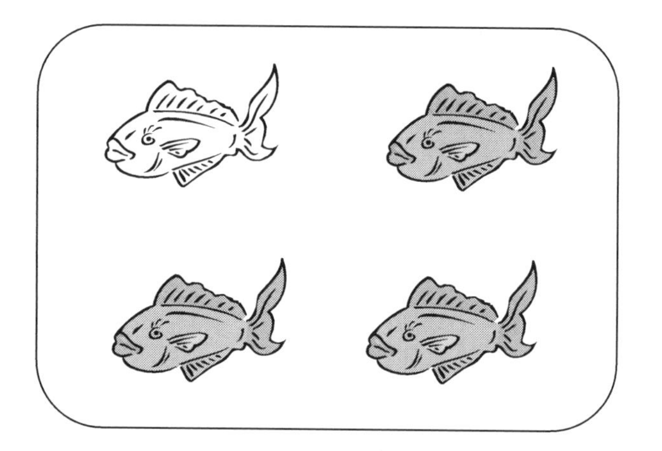

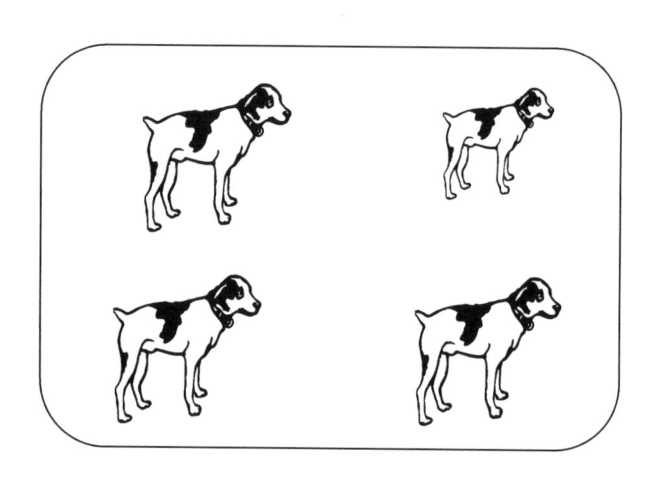

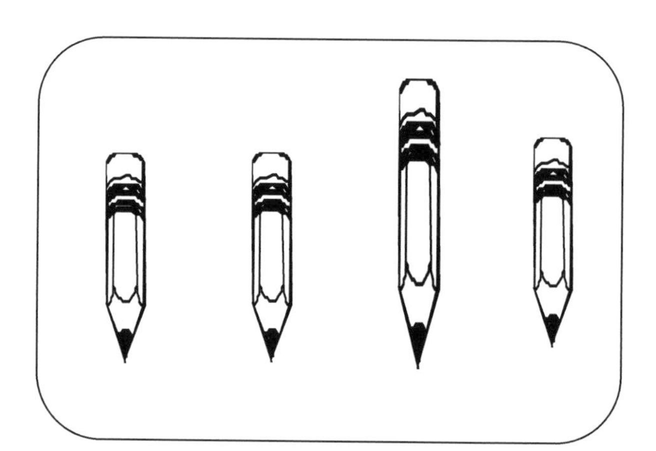

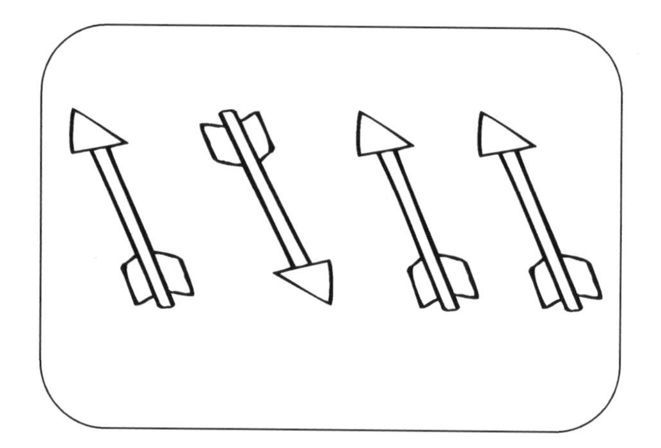

✵In each line, circle and then colour the picture that is different.

✱ In each box, circle and then colour the picture that is bigger.

✱ Draw a picture that is bigger than the one in the box.

✵ In each box, circle and then colour the object that is smaller.

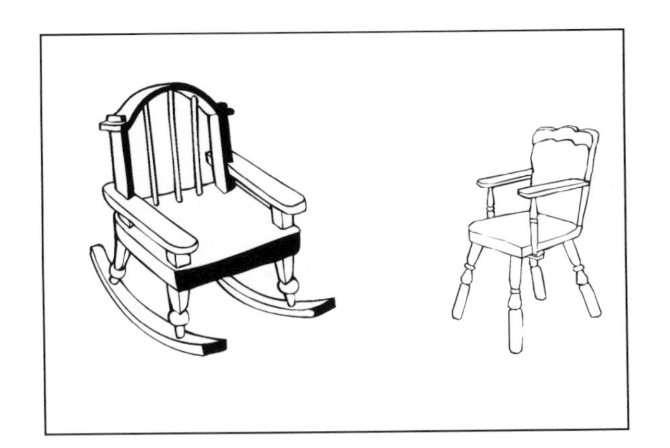

✵ Draw a picture that is smaller than the one in the box.

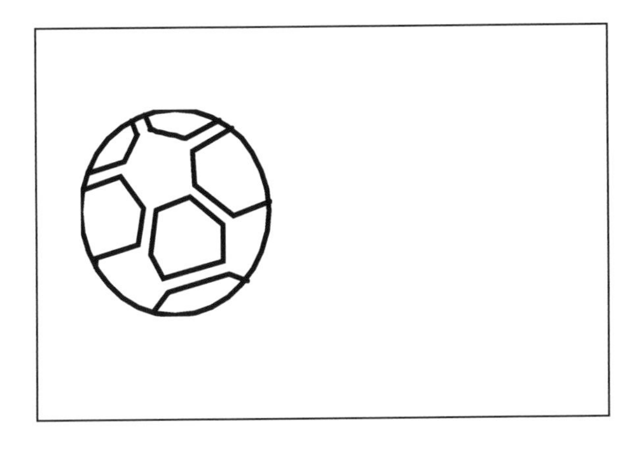

✱Draw the next picture to complete each pattern.

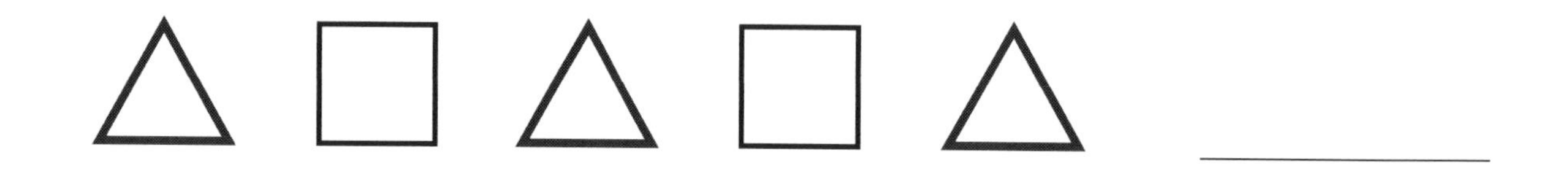

c o c o c ______

✷Draw the pictures to complete each pattern.

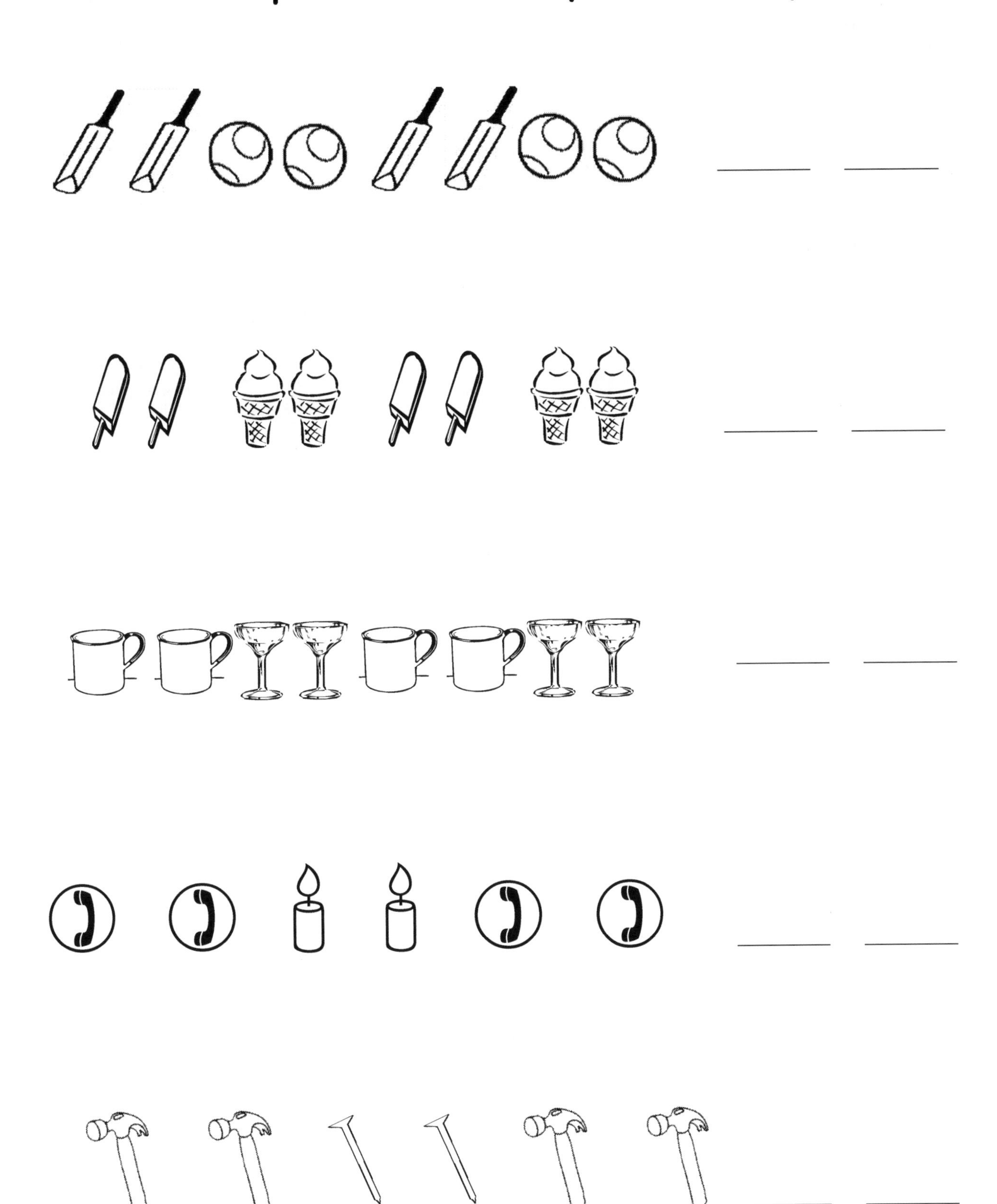

* In each box, circle and then colour the objects that are full.

* In each box, circle and then colour the objects that are empty.

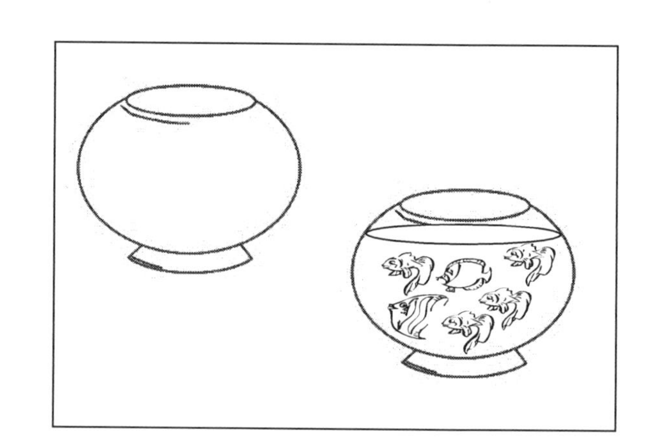

✵In each box, circle and then colour the object that is lighter.

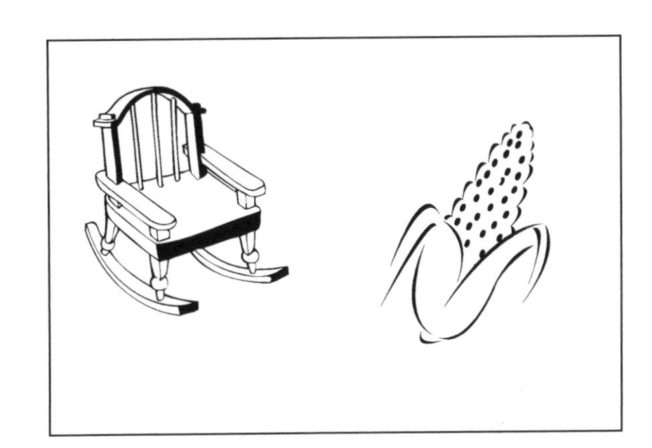

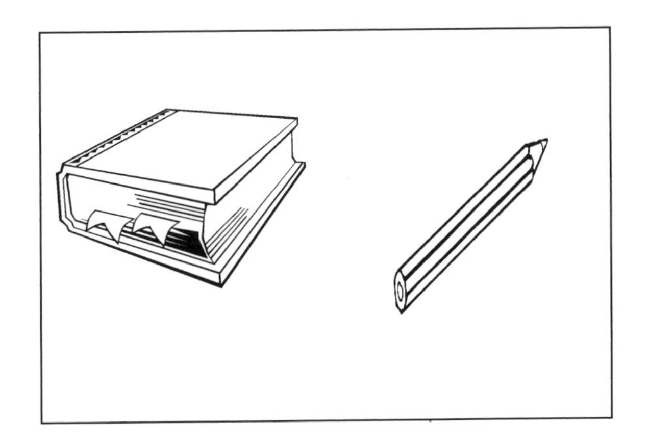

✱In each box, circle and then colour the object that is heavier.

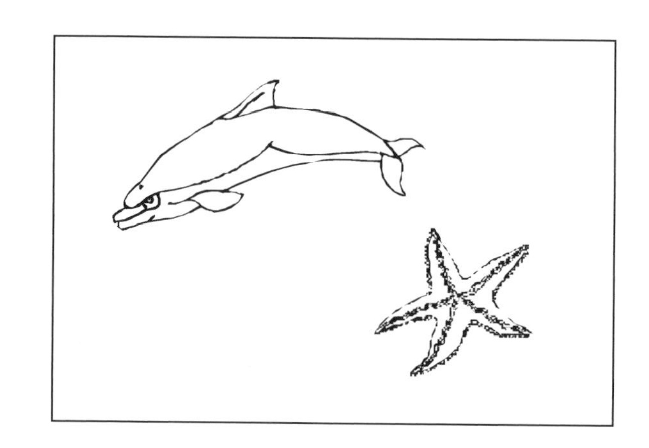

✷In each box, circle and then colour the object that is taller.

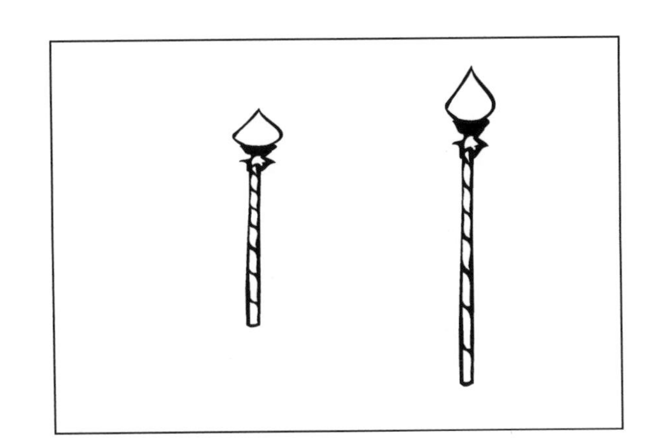

✷Draw a picture that is taller than the one in the box.

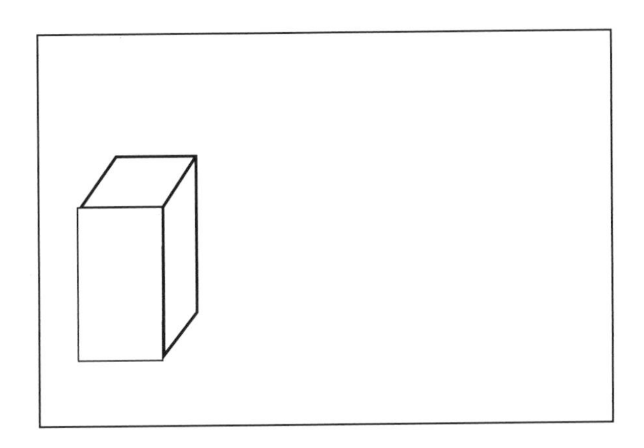

* In each box, circle and then colour the object that is shorter.

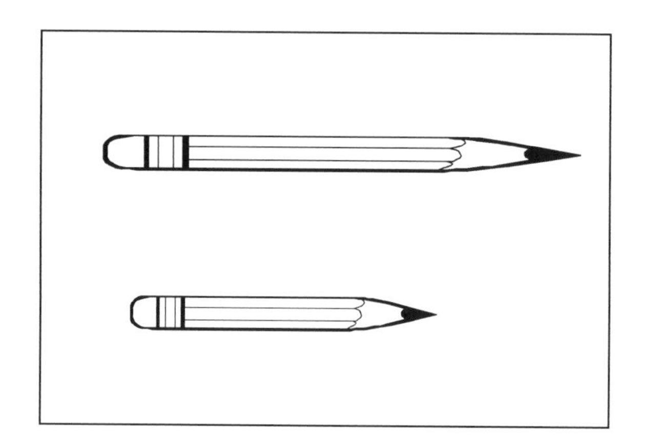

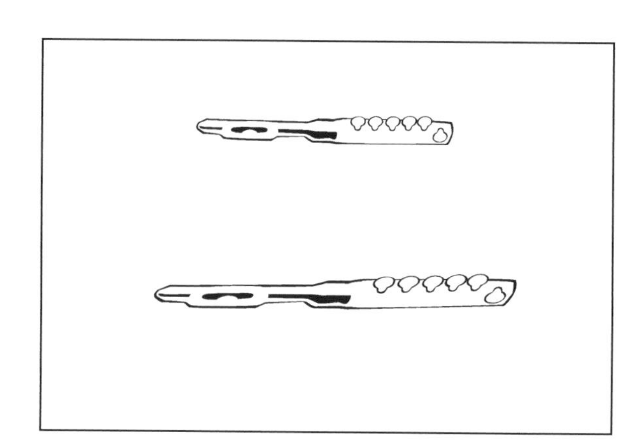

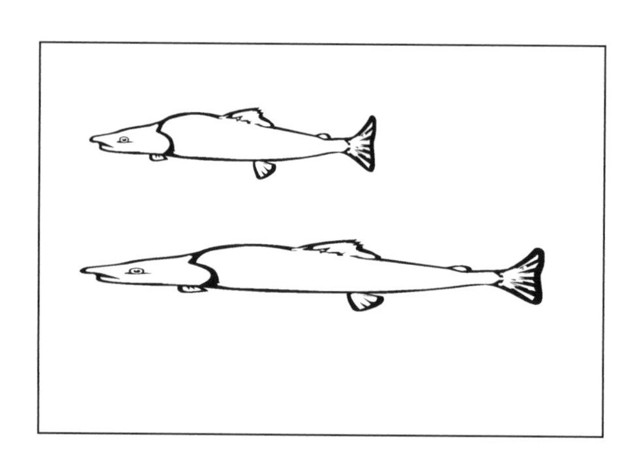

* Draw a picture that is shorter than the one in the box.

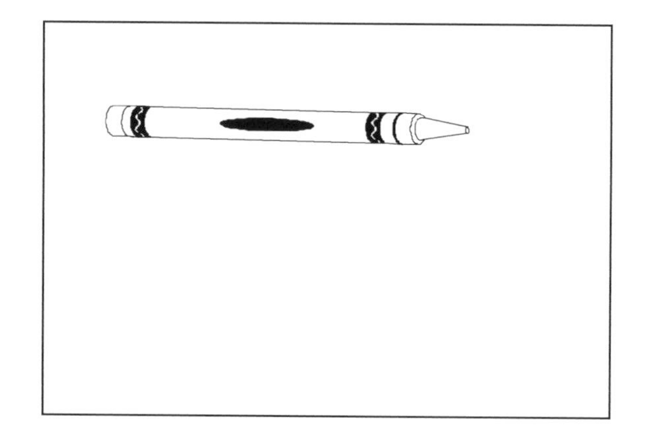

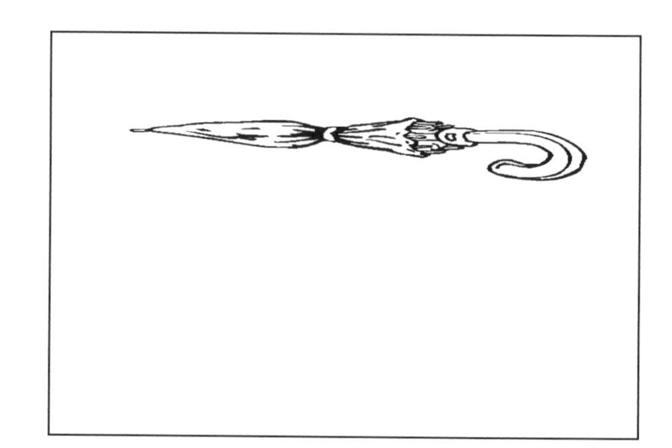

* In each box, circle and then colour the object that is longer.

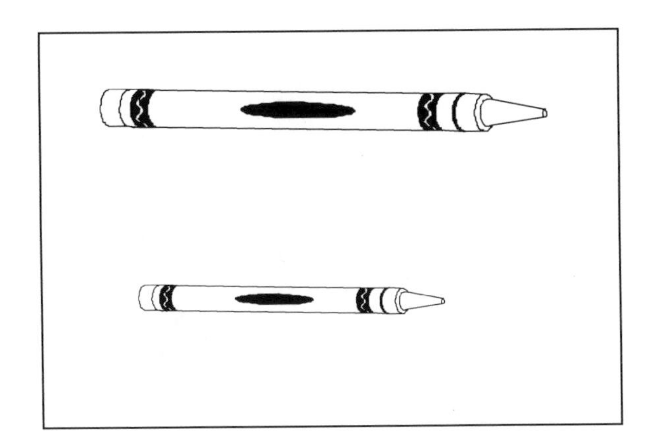

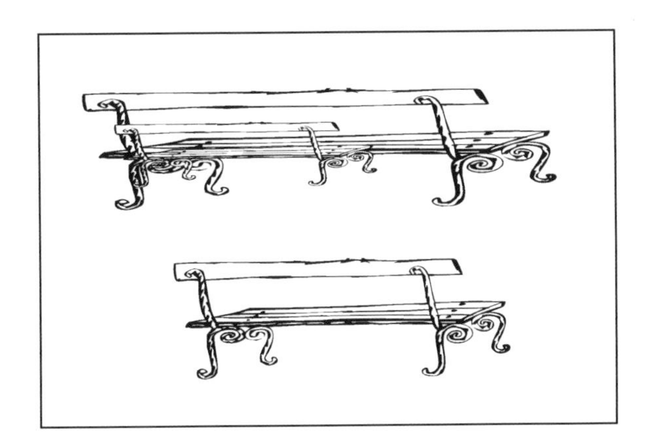

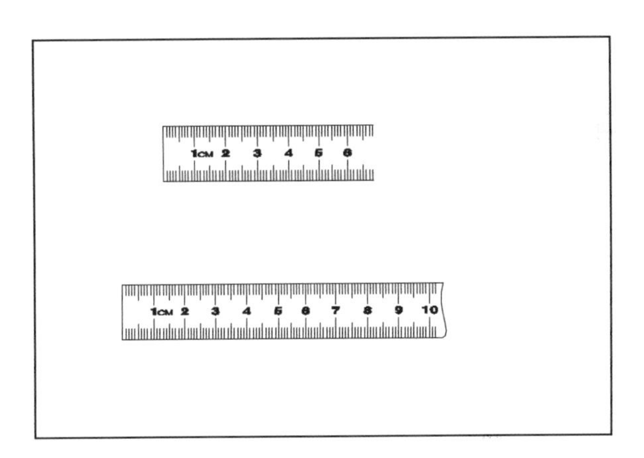

* Draw a picture that is longer than the one in the box.

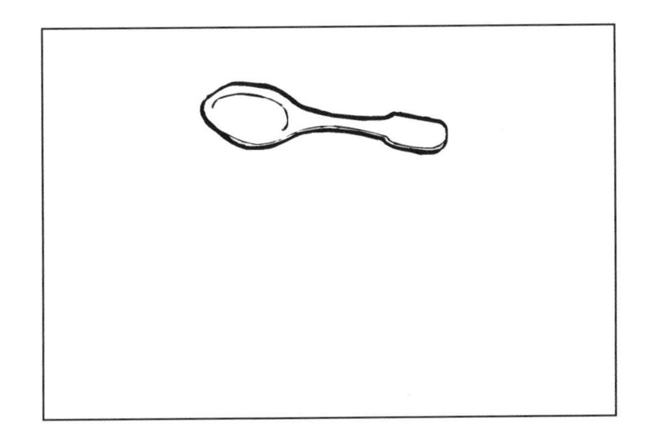

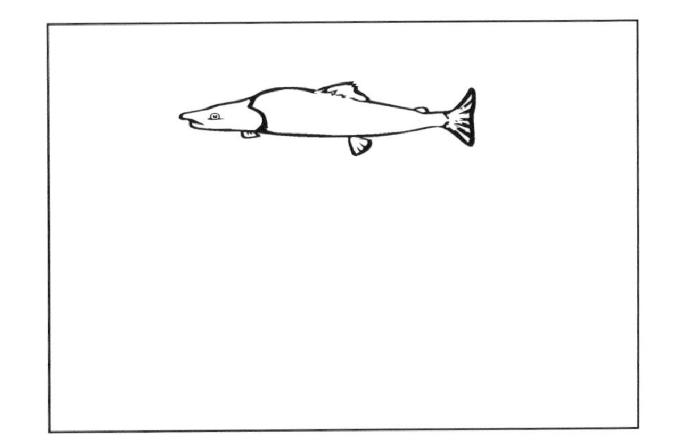

✱ In each box, circle and then colour the object that holds more.

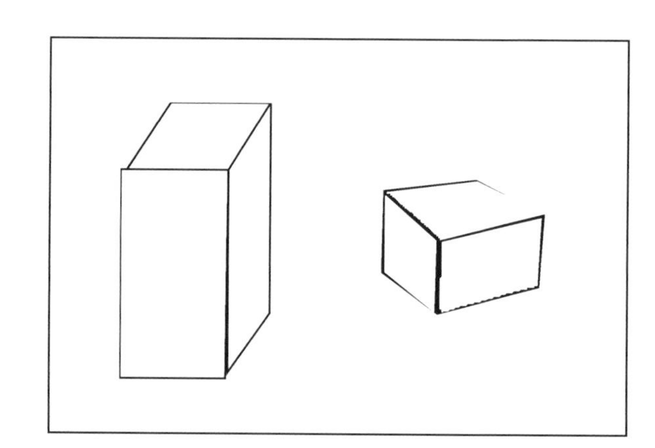

✱ Draw an object that holds more than the one in the box.

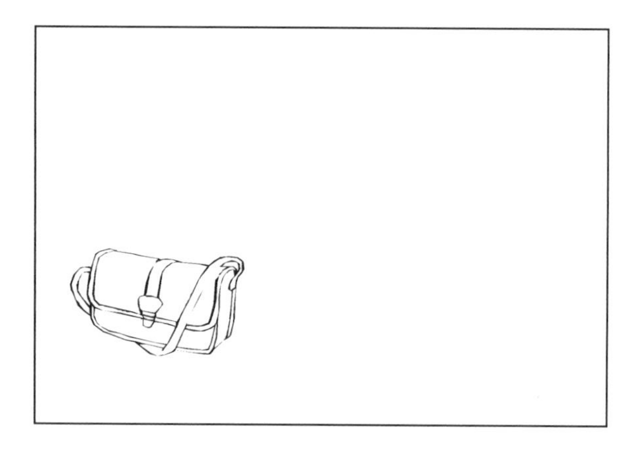

✲ In each box, circle and then colour the object that holds less.

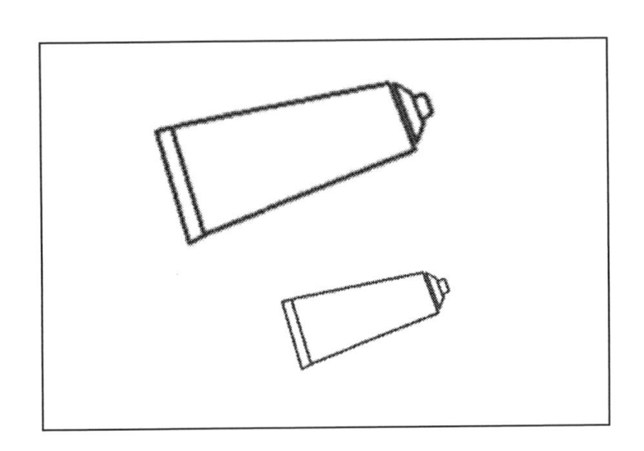

✲ Draw an object that holds less than the one in the box.

✵Draw the pictures to complete each pattern.

Part 2

✷Count the object. Read the number name and then colour the picture.

✱Trace the ones.

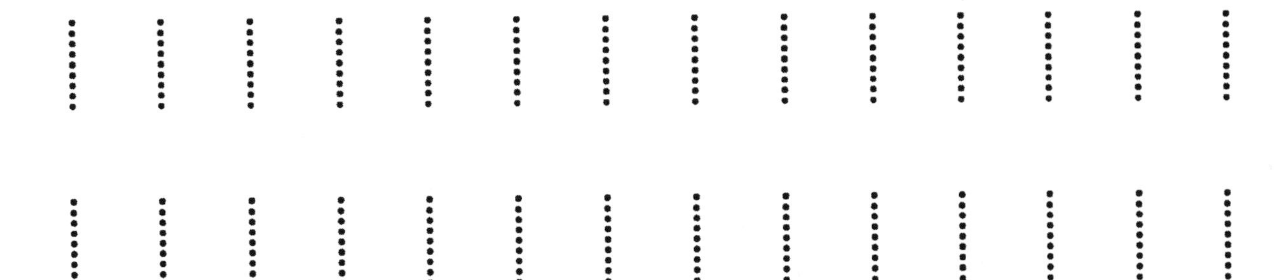

✱Circle one object in each set.

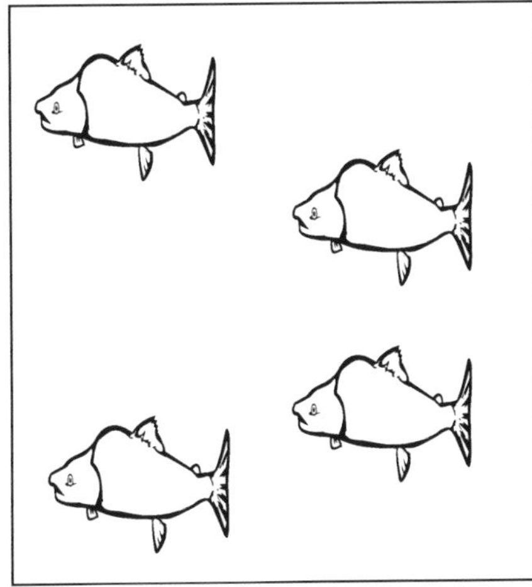

✷Write some ones on the line. Start at the dot.

1

✷Draw objects for the number.

1 ball

1 cat

1 balloon

1 box

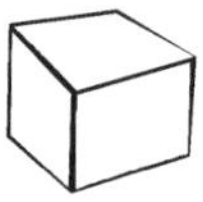

1 sun

1 fish

✵Circle and then colour the sets that are one.

✵Count the objects. Read the number name and then colour the pictures.

✱Trace the twos.

2 2 2 2 2 2 2 2 2 2

2 2 2 2 2 2 2 2 2 2

✱Circle two objects in each set.

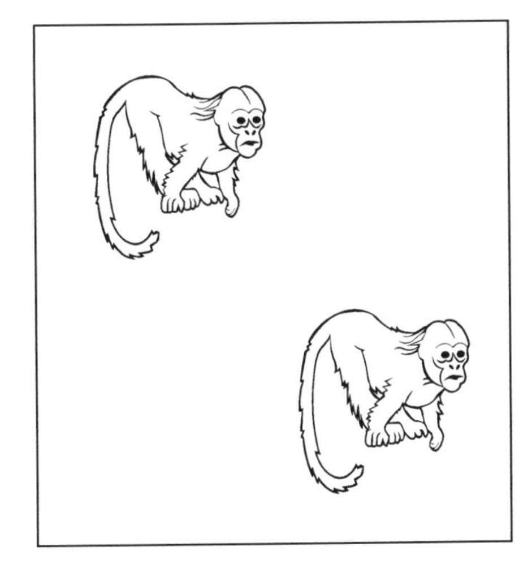

✱Write some twos on the line. Start at the dot.

2

✱Draw objects for the number.

2 balls

2 cups

2 kites

2 bags

2 eggs

2 apples

✱Circle and then colour the sets that are two.

✲Complete the sets to make them two.

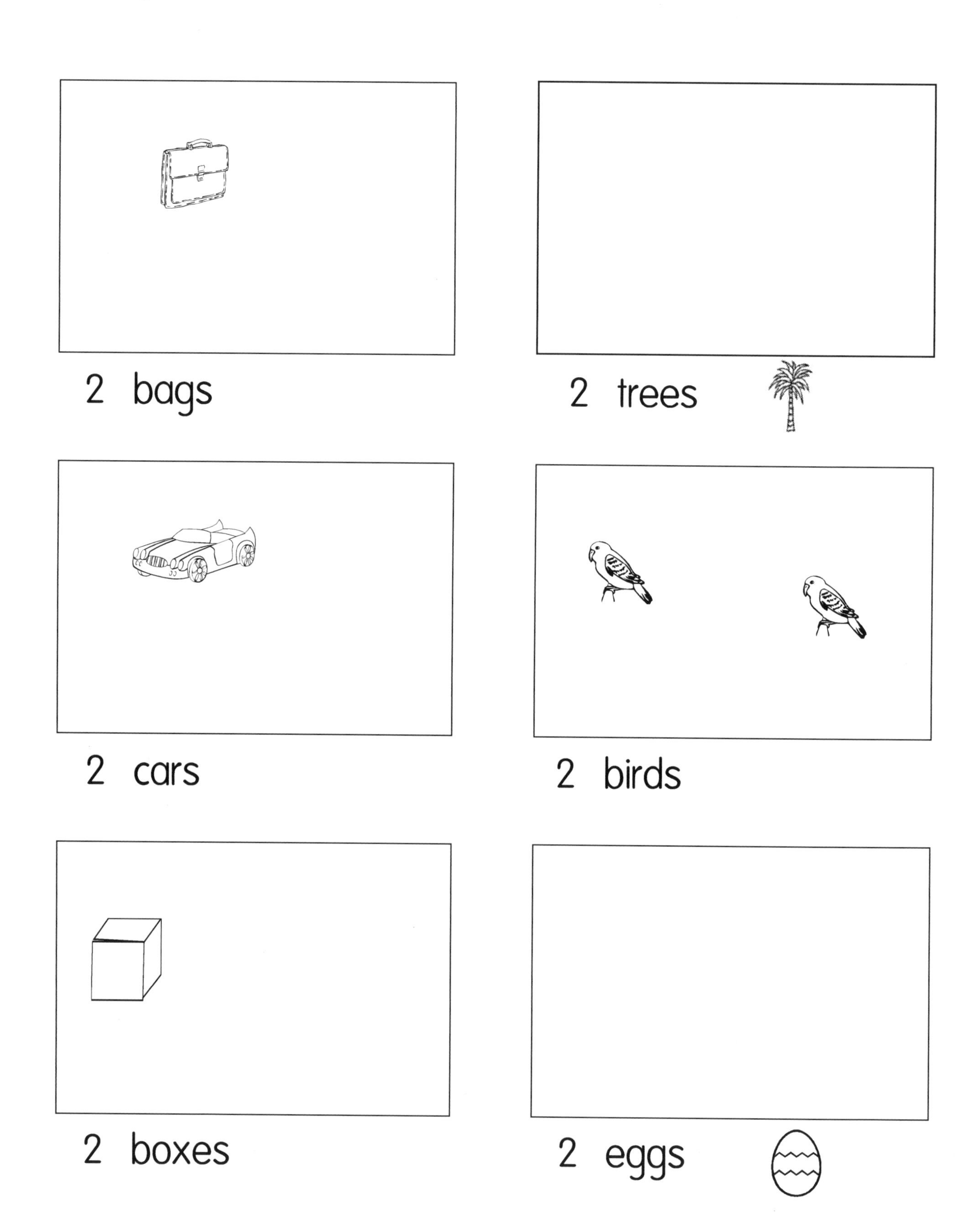

✵ Count the objects. Read the number name and then colour the pictures.

✱Trace the threes.

3 3 3 3 3 3 3 3 3 3

3 3 3 3 3 3 3 3 3 3

✱Circle three objects in each set.

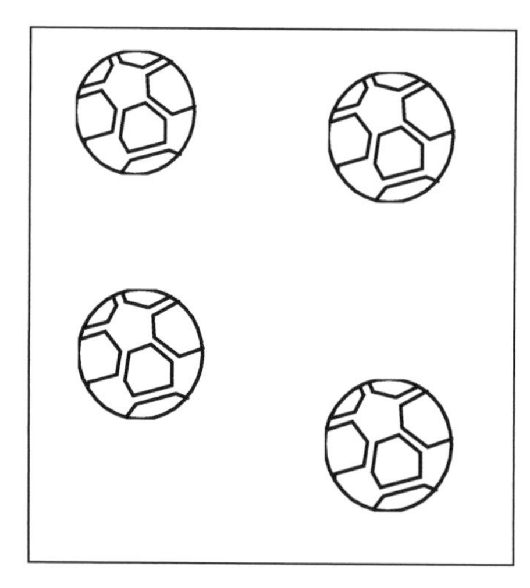

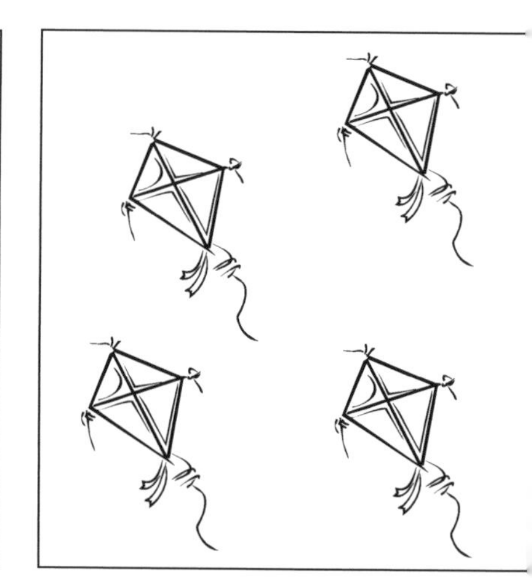

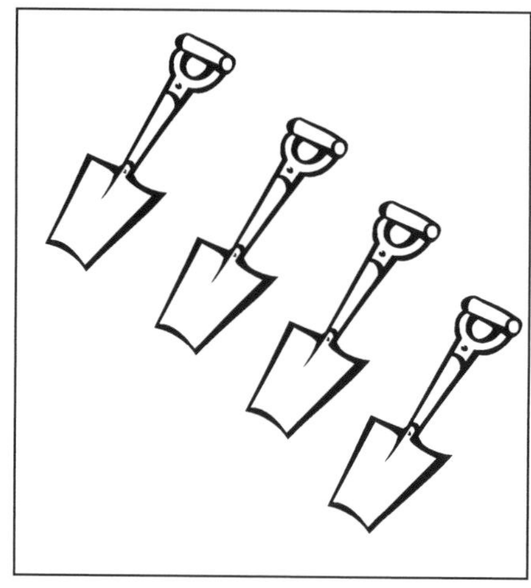

✻Write some threes on the line. Start at the dot.

3 . . . .

✻Draw objects for the number.

3 circles

3 cars

3 lollipops

3 boys

3 eggs

3 trees

✷ Circle and then colour the sets that are three.

✷Complete the sets to make them three.

✱Count the objects. Read the number name and then colour the pictures.

✸Trace the fours.

4 4 4 4 4 4 4 4 4 4

4 4 4 4 4 4 4 4 4 4

✸Circle four objects in each set.

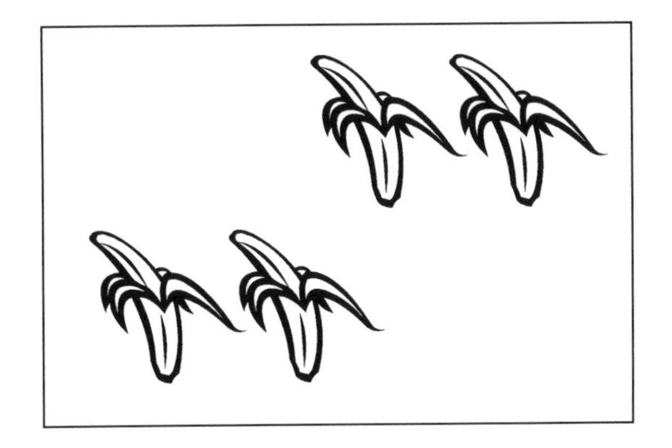

✱Write some fours on the line. Start at the dot.

4

✱Draw objects for the number.

4 balls

4 pencils

4 books

4 boxes

4 glasses

4 bananas

✲Circle and then colour the sets that are four.

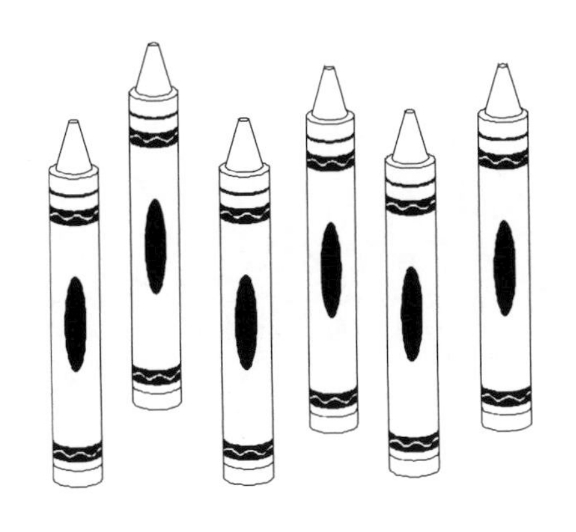

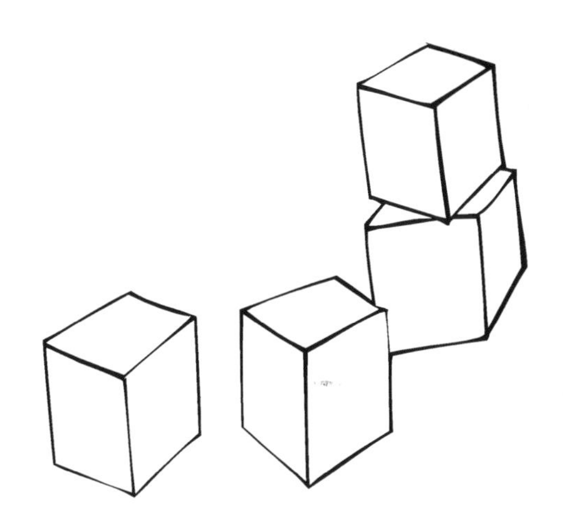

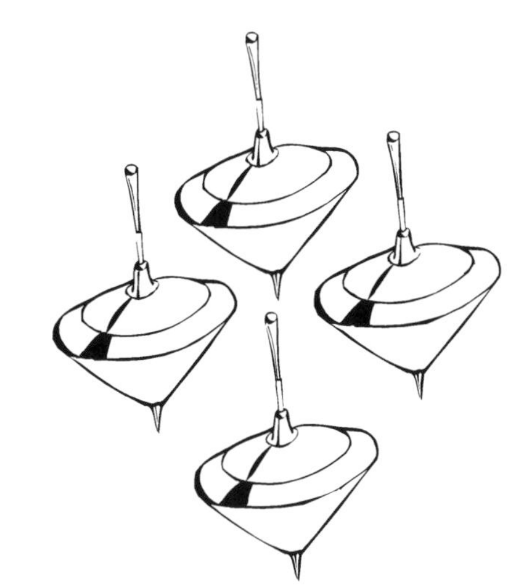

✵Complete the sets to make them four.

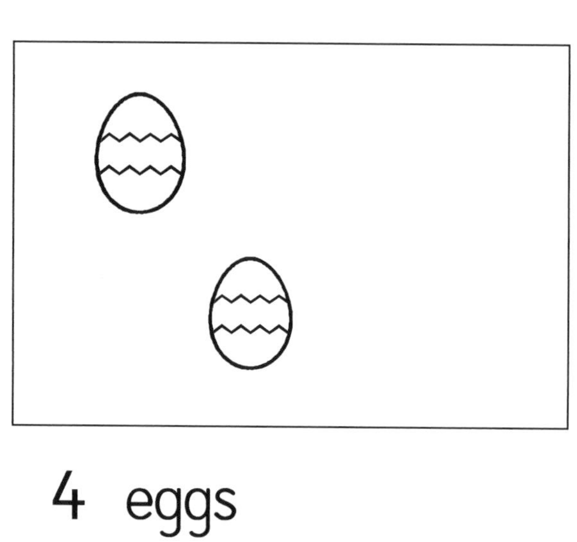

4 eggs

4 trees

4 pears

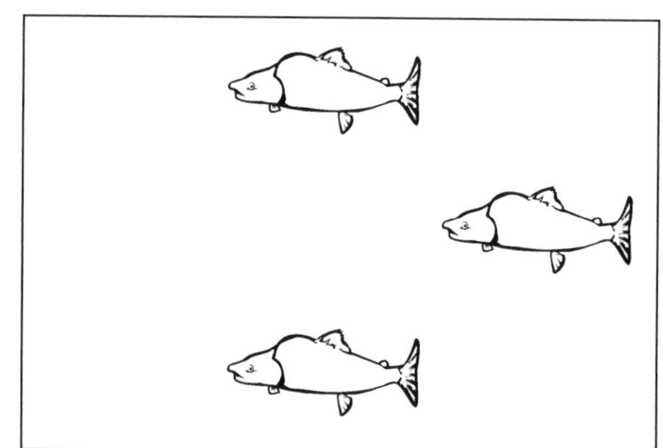

4 fishes

4 boys

4 cups

✶Count the objects. Read the number name and then colour the pictures.

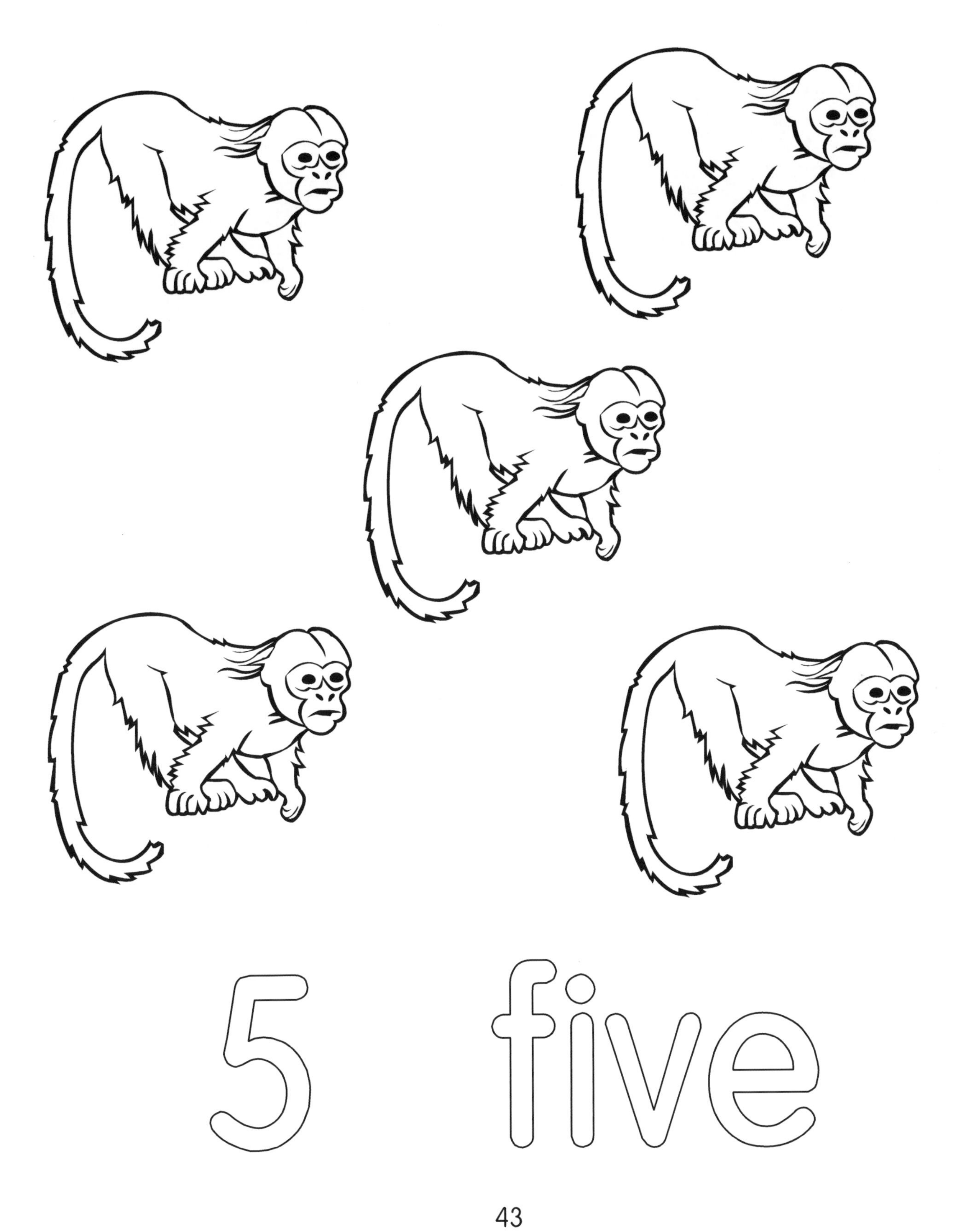

5 five

✱Trace the fives.

5 5 5 5 5 5 5 5 5 5

5 5 5 5 5 5 5 5 5 5

✱Circle five objects in each set.

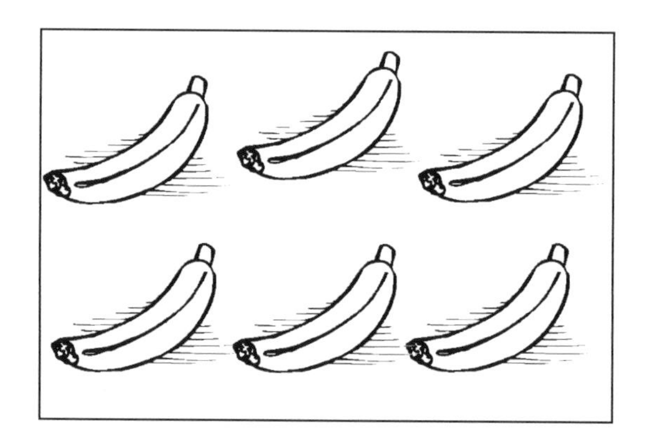

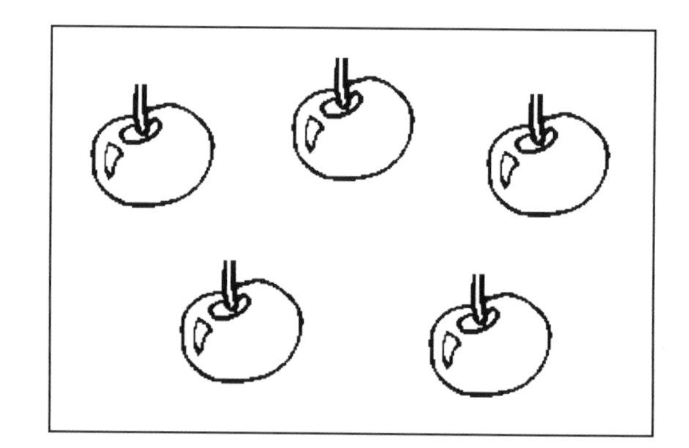

✻Write some fives on the line. Start at the dot.

5 . . . .

✻Draw objects for the number.

5 cups 

5 fishes 

5 pencils 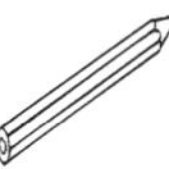

5 bananas

5 eggs 

5 apples 

✱Circle and then colour the sets that are five.

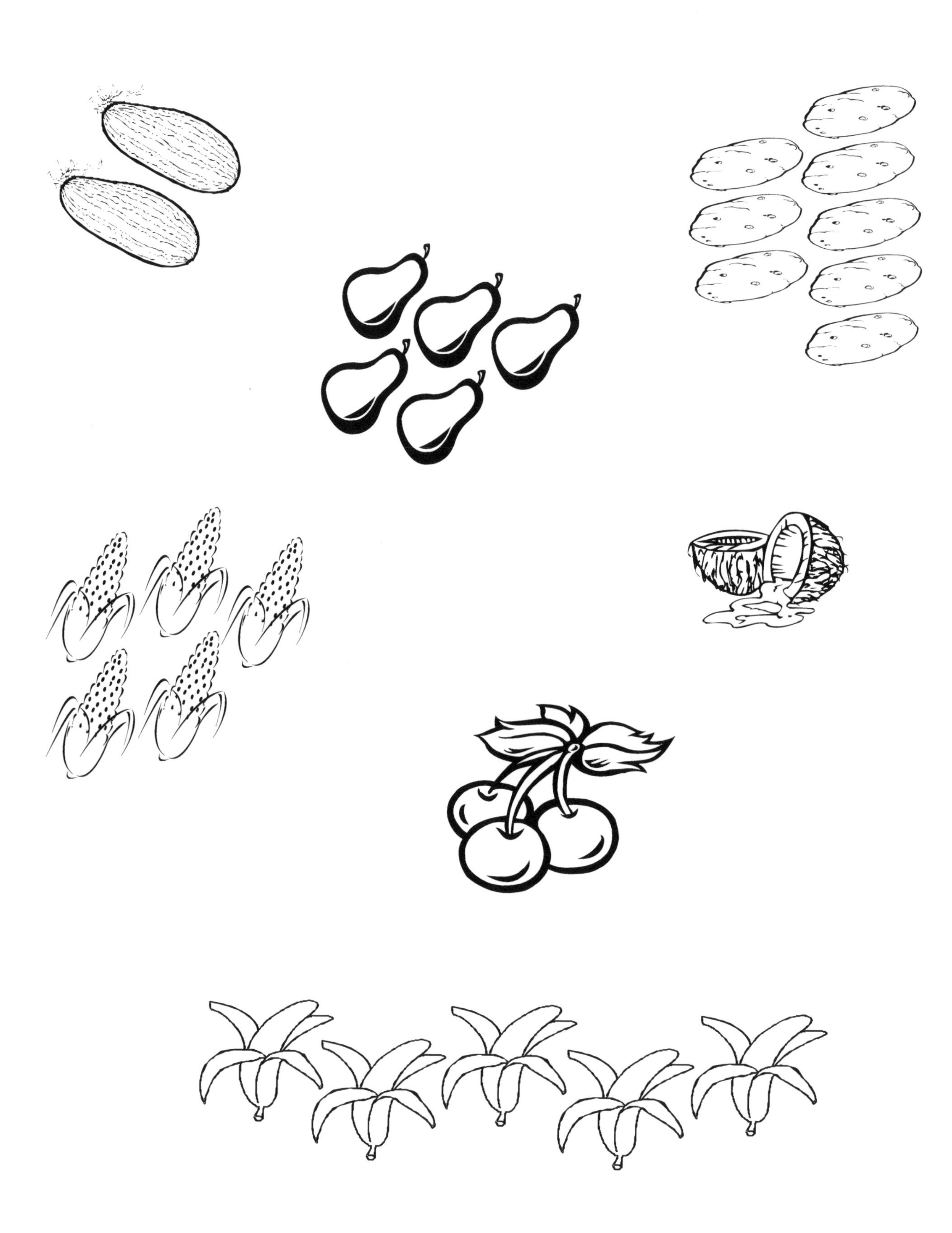

✷Complete the sets to make them five.

✷Trace the zeros.

0 0 0 0 0 0 0 0 0

✷Read and then colour the number and the number name.

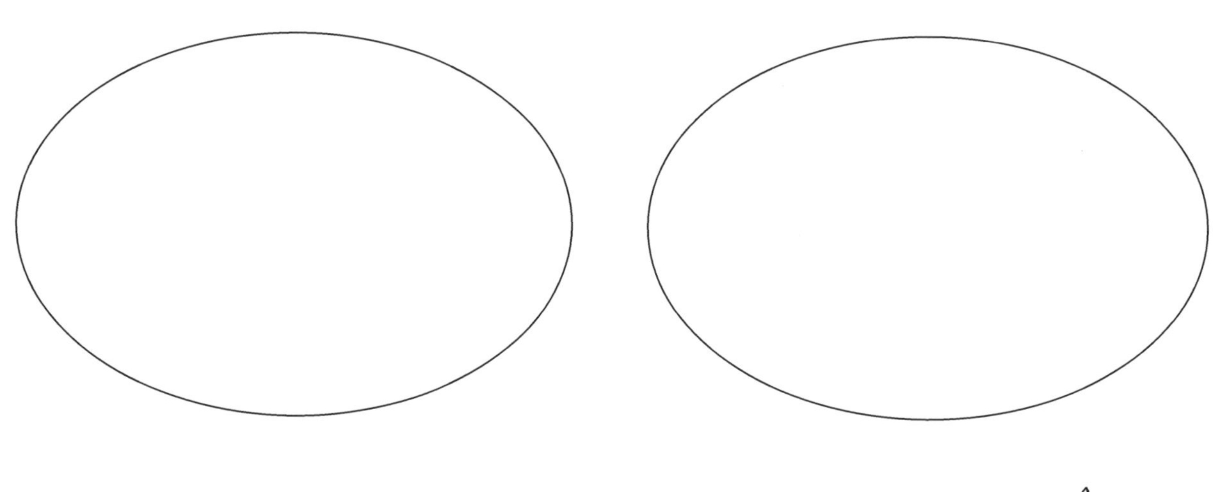

0 houses

0 rabbits 

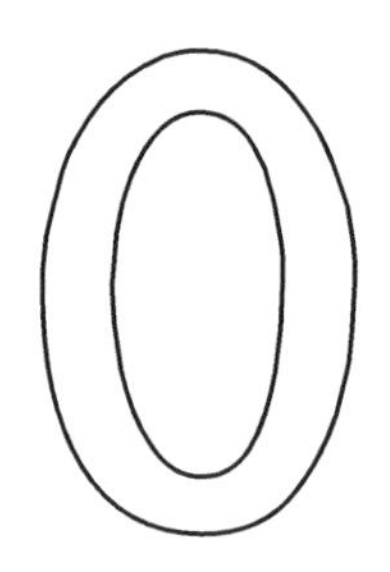 zero

✱Write some zeros. Start at the dot.

0 . . . .

✱Draw objects in the ovals.

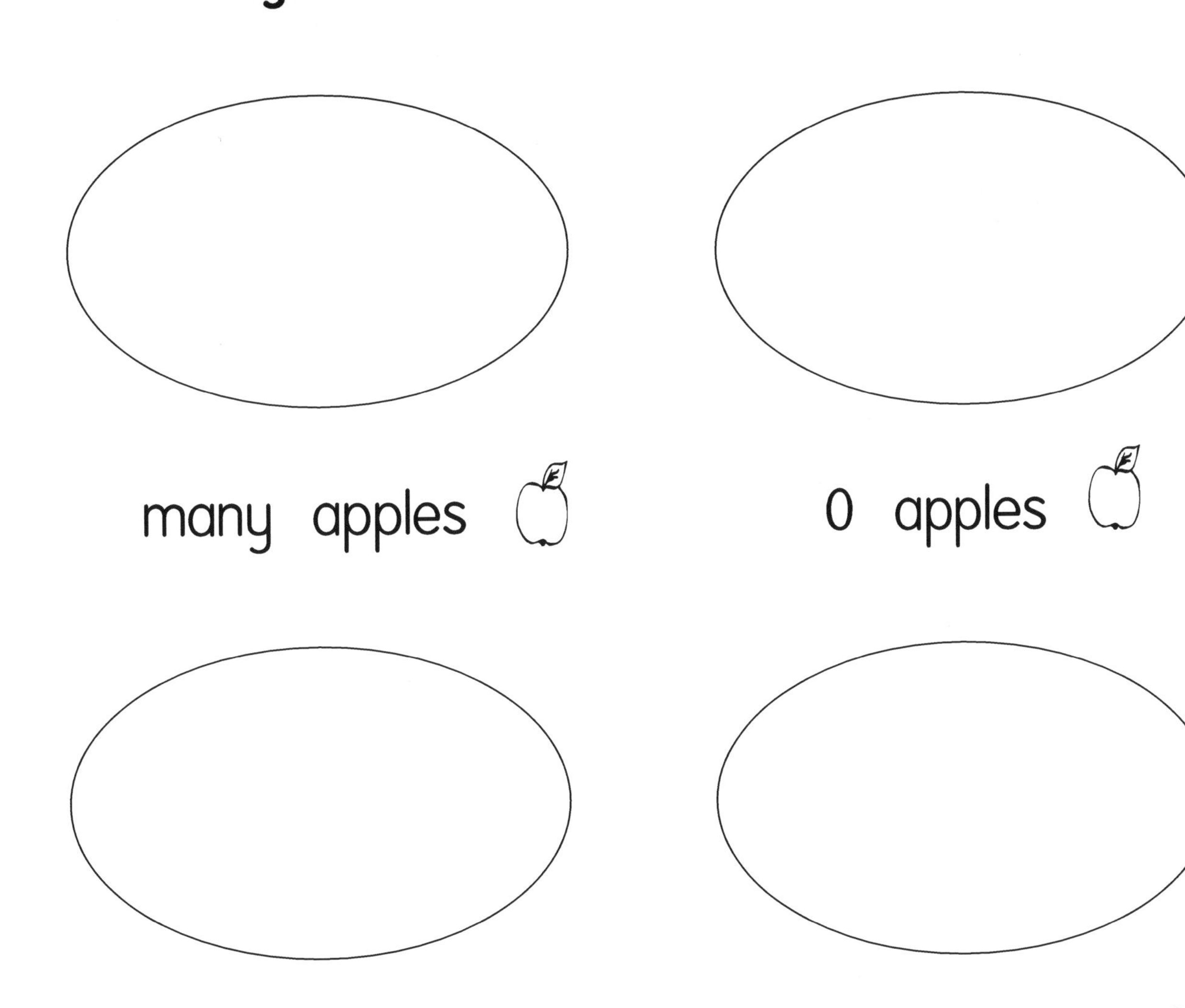

many apples

0 apples

a few balloons 

0 balloons 

✷Circle the correct amount.

a) Circle two (2) balls.

b) Circle one (1) pig.

c) Circle five (5) cars.

 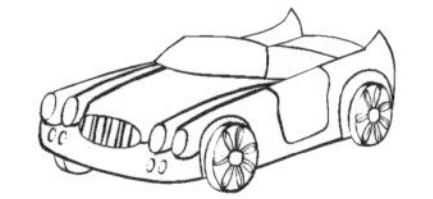 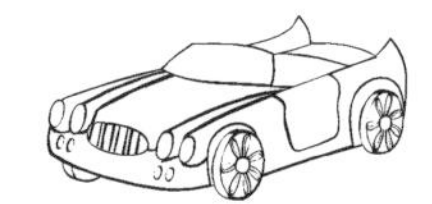 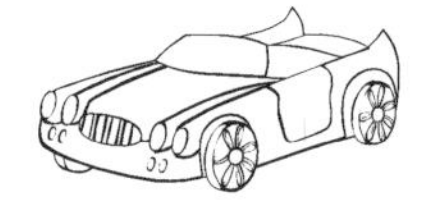 

d) Circle three (3) pears.

✱Draw the objects for each number.

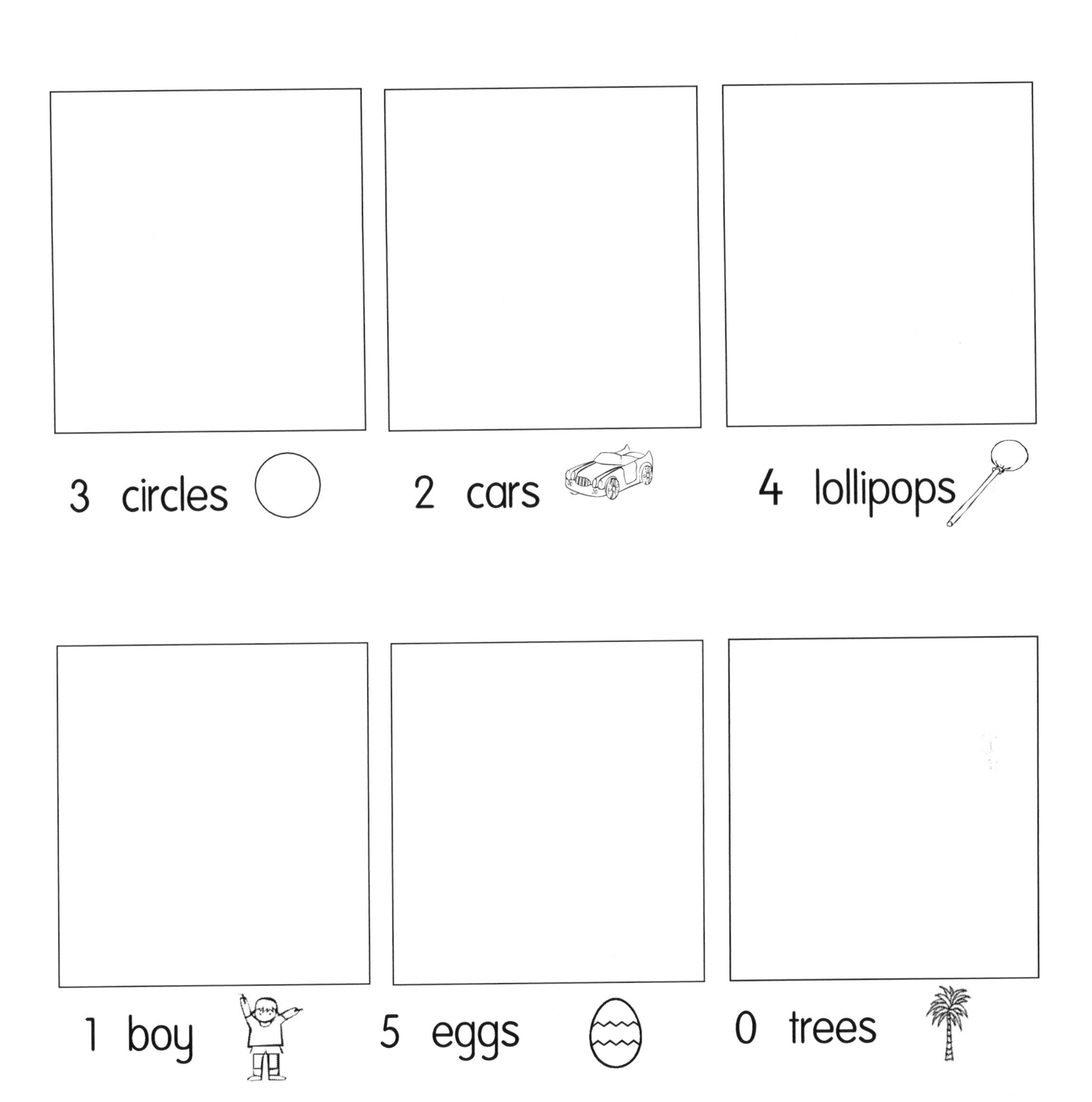

✱ Complete the sets.

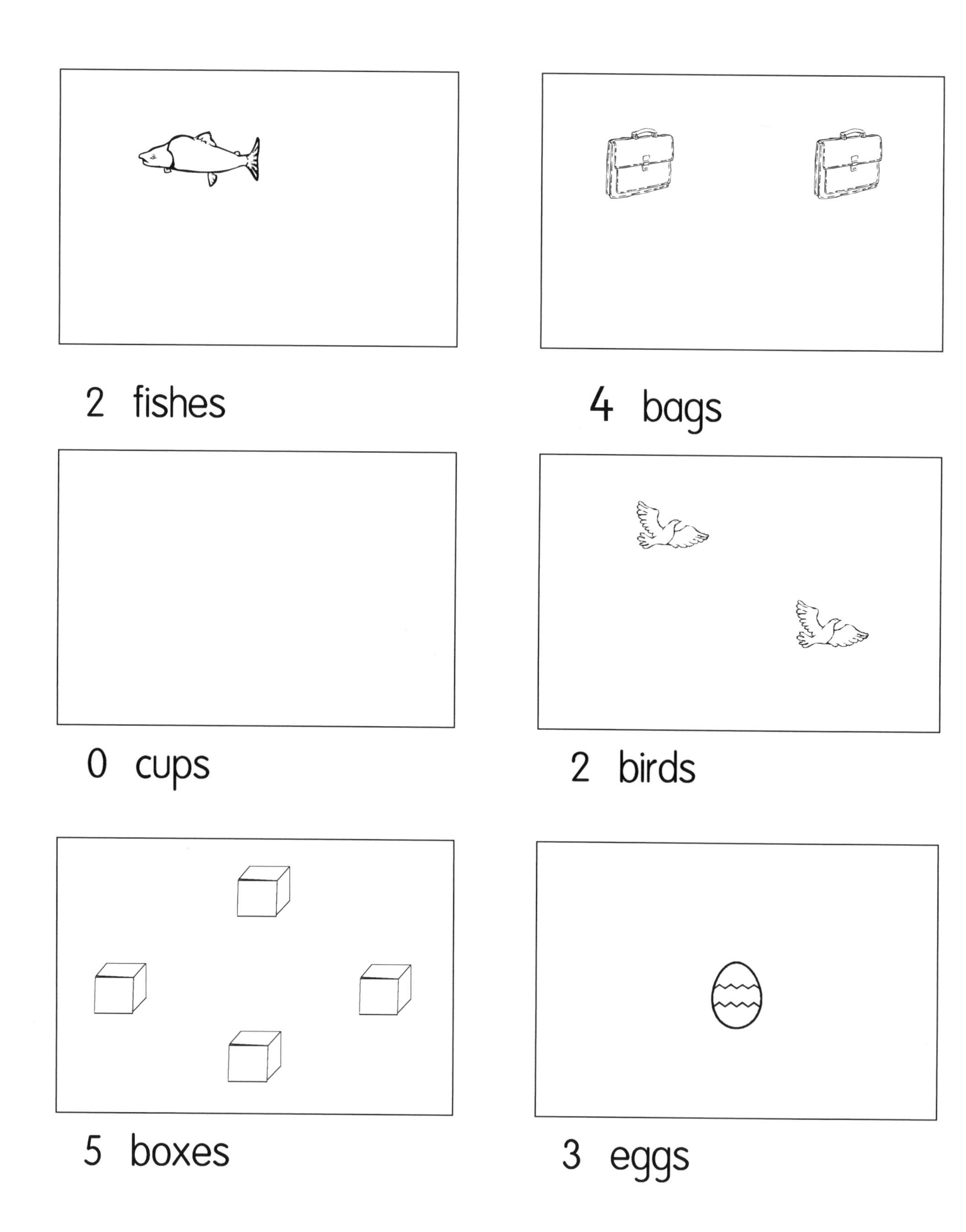

✲Count the objects and match them to the number.

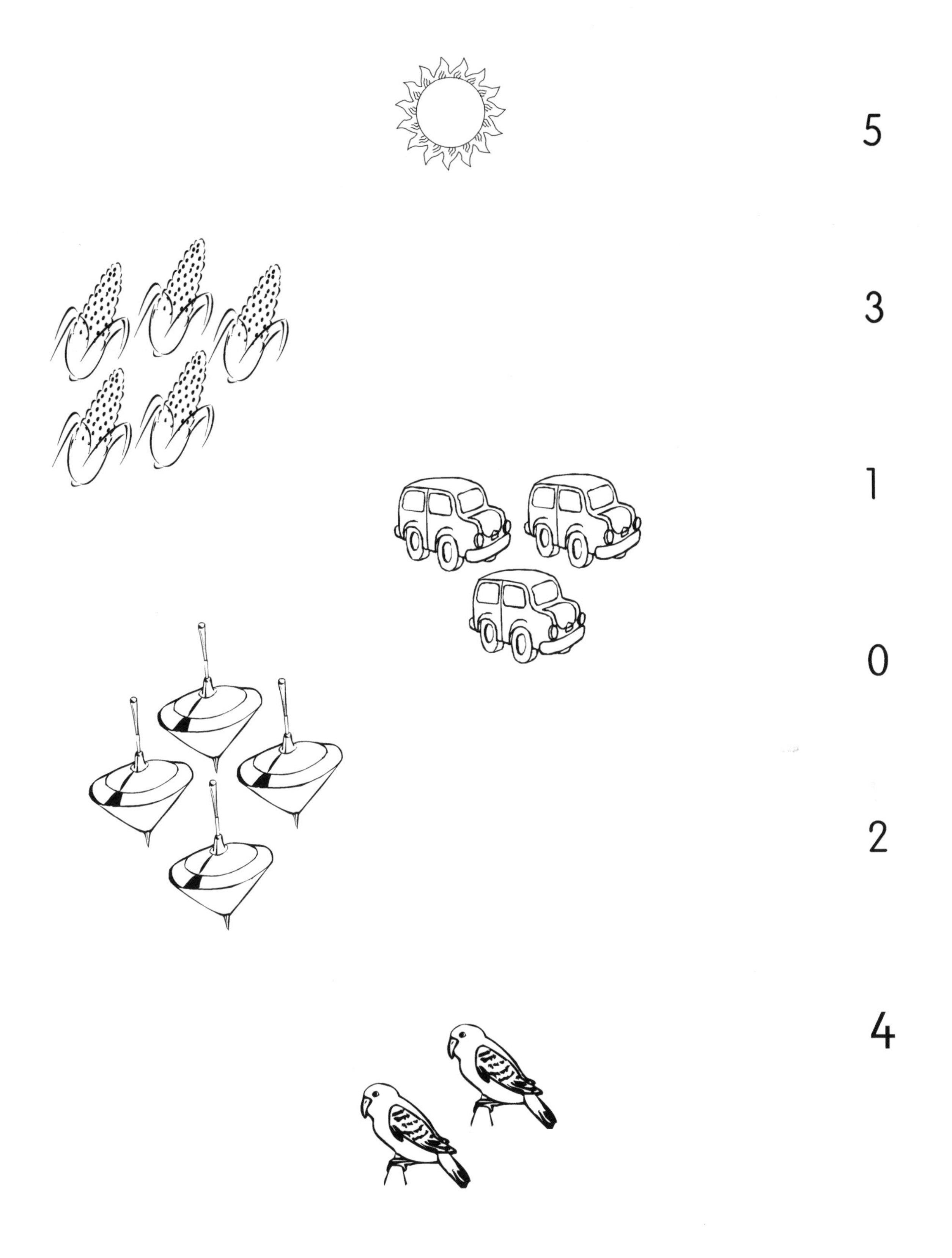

✲ Count the objects and then circle the correct number in the box.

5	4	8

2	3	5

7	5	1

5	6	4

2	3	0

✻Count the objects. Read the number name and then colour the pictures.

✷Trace the sixes.

6 6 6 6 6 6 6 6 6 6

6 6 6 6 6 6 6 6 6 6

✷Circle six objects in each set.

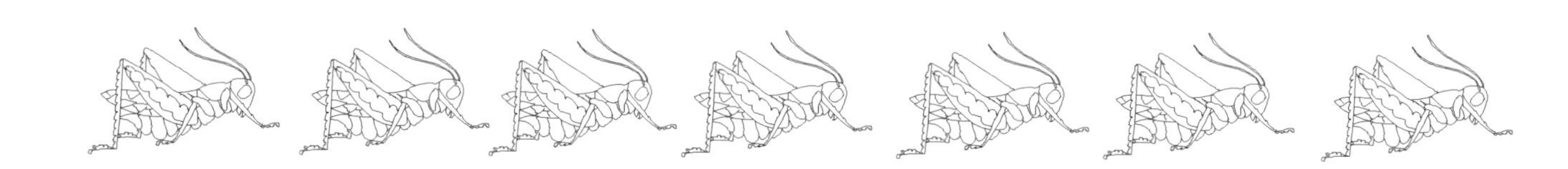

✱Write some sixes on the line. Start at the dot.

6 . . . .

✱Draw objects in the boxes for the number.

6 hats | 6 balloons | 6 cakes

6 bees | 6 potatoes | 6 bells

✷ Circle and then colour the sets that are six.

✵Complete the sets to make them six.

6 cakes

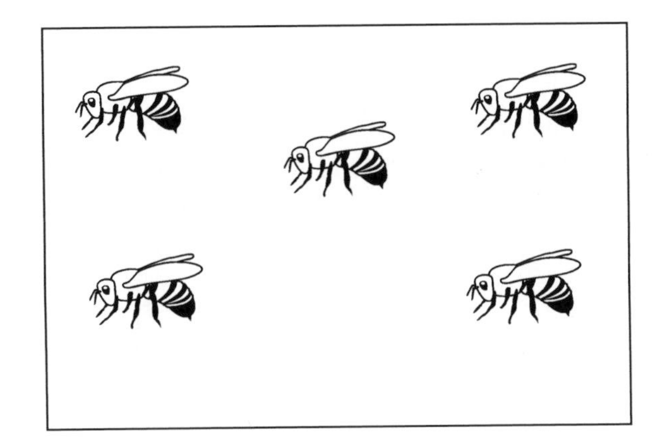

6 bees

6 glasses

6 fishes

6 balloons

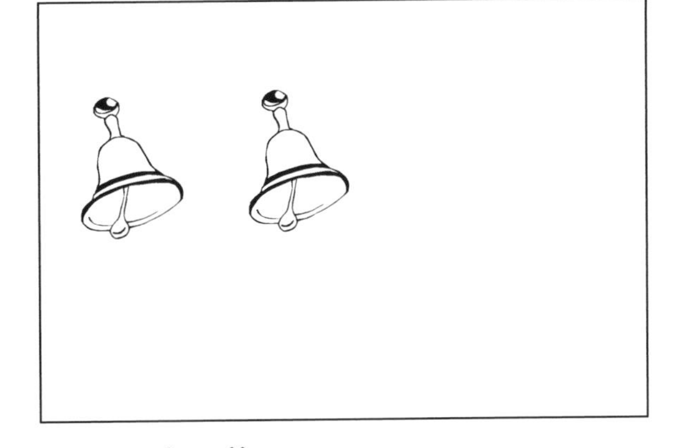

6 bells

✱ Count the objects. Read the number name and then colour the pictures.

✱Trace the sevens.

7 7 7 7 7 7 7 7 7 7

7 7 7 7 7 7 7 7 7 7

✱Circle seven objects in each set.

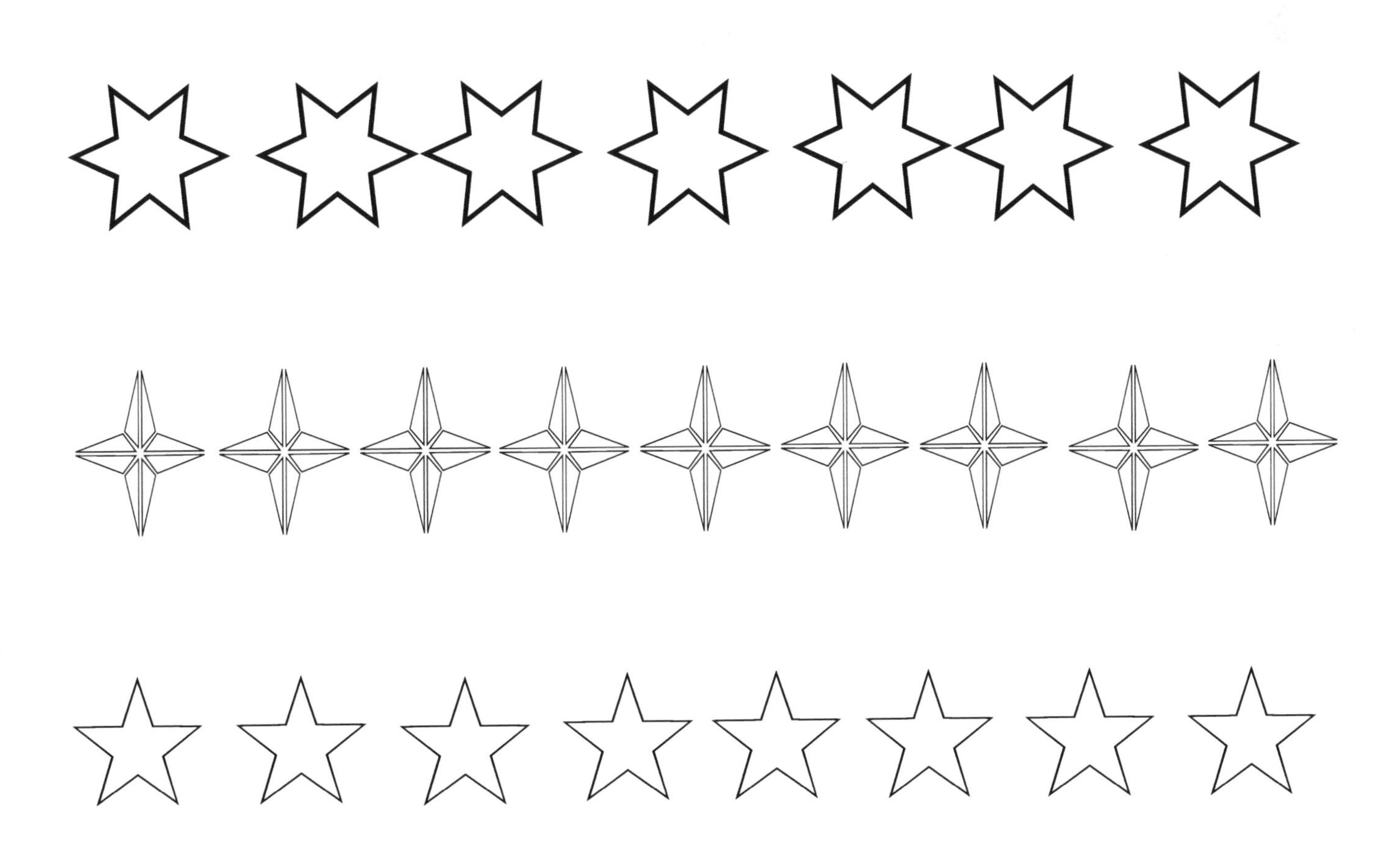

*Write some sevens on the line. Start at the dot.

7 . . . .

*Draw objects in the boxes for the number.

7 bells	7 glasses	7 triangles △
7 cakes	7 spades	7 cups

✶ Circle and then colour the sets that are seven.

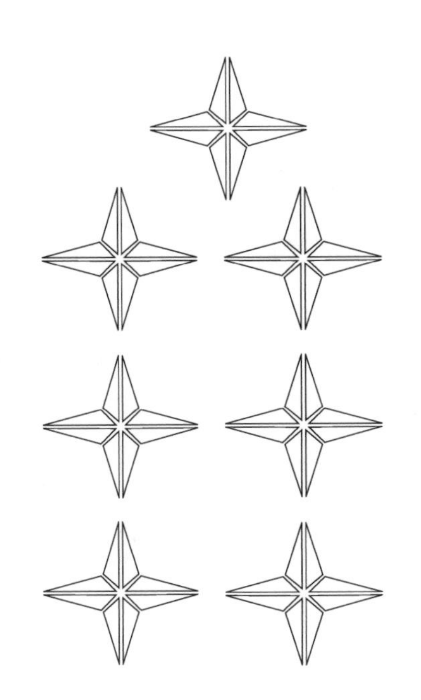

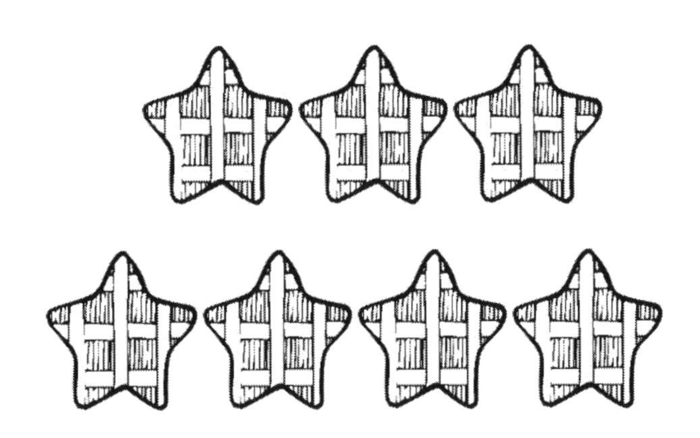

✳Complete the sets to make them seven.

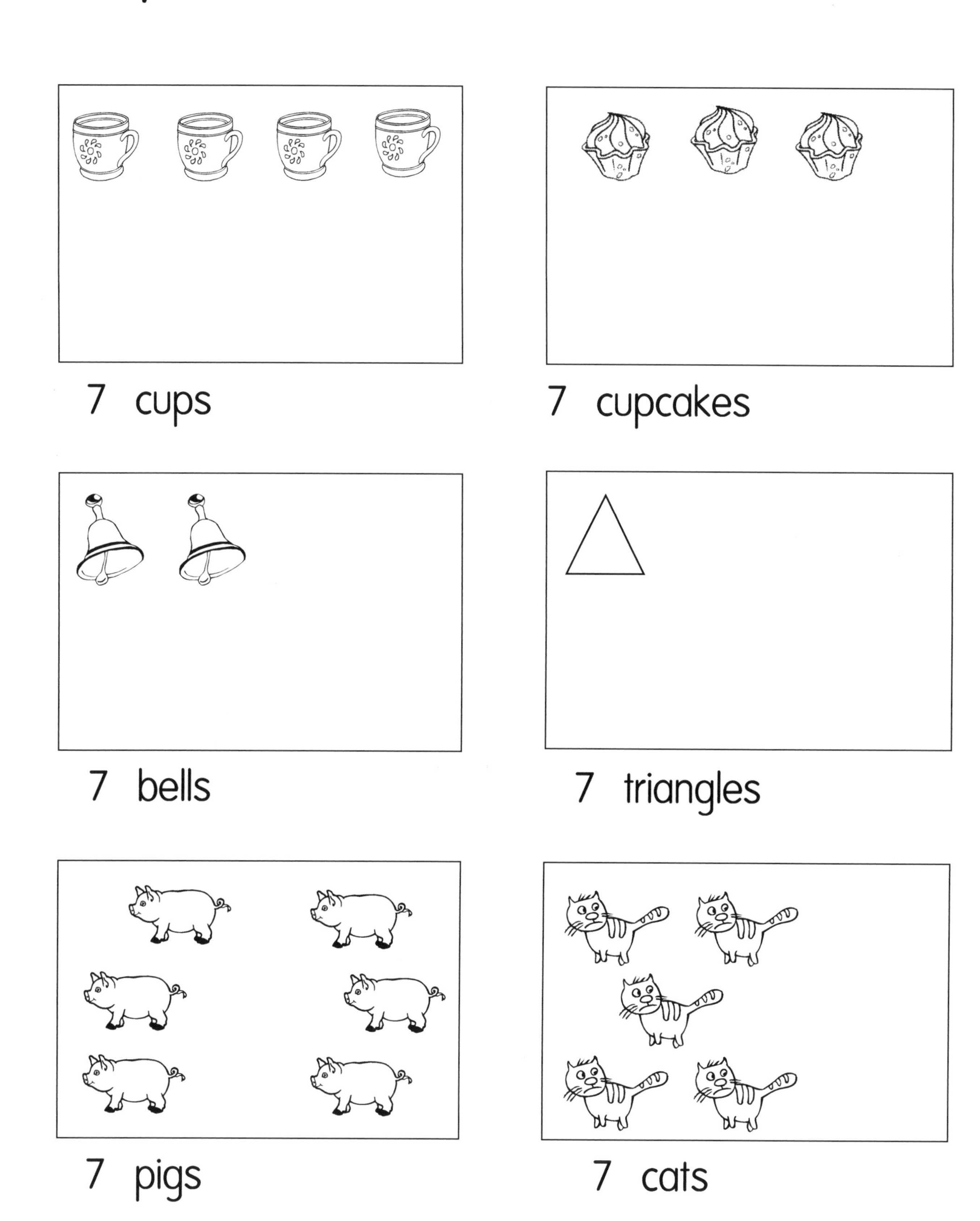

✷Count the objects. Read the number name and then colour the pictures.

✵Trace the eights.

8 8 8 8 8 8 8 8 8 8

8 8 8 8 8 8 8 8 8 8

✵Circle eight objects in each set.

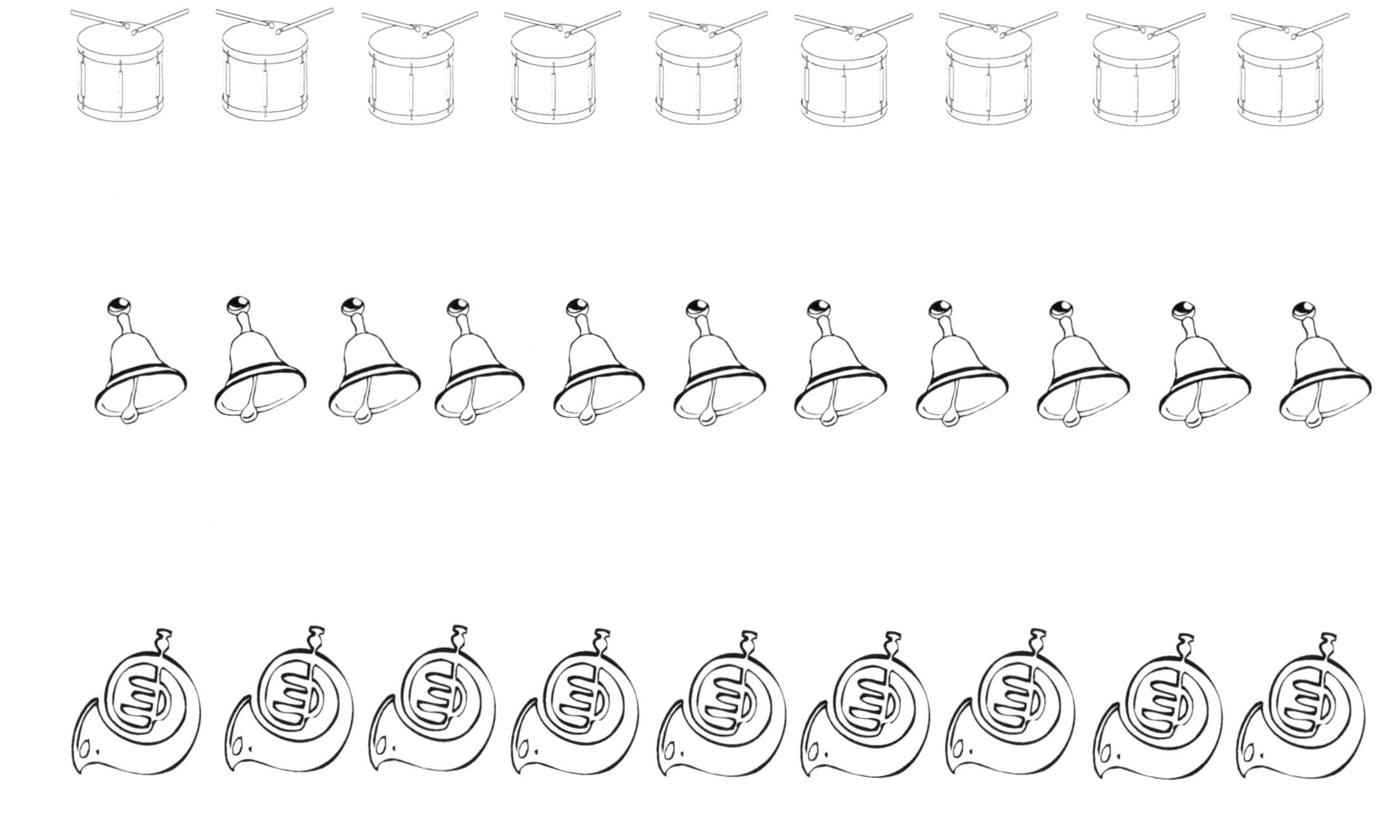

✱Write some eights on the line. Start at the dot.

8 . . . .

✱Draw objects in the boxes for the number.

8 hats	8 circles ○	8 triangles △
8 balls	8 pencils	8 bells

✱ Circle and then colour the sets that are eight.

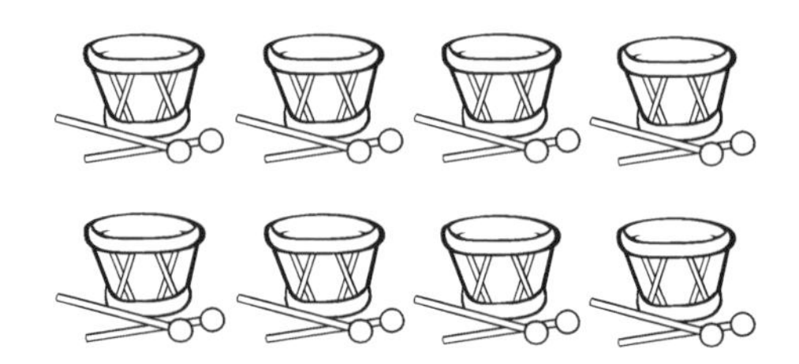

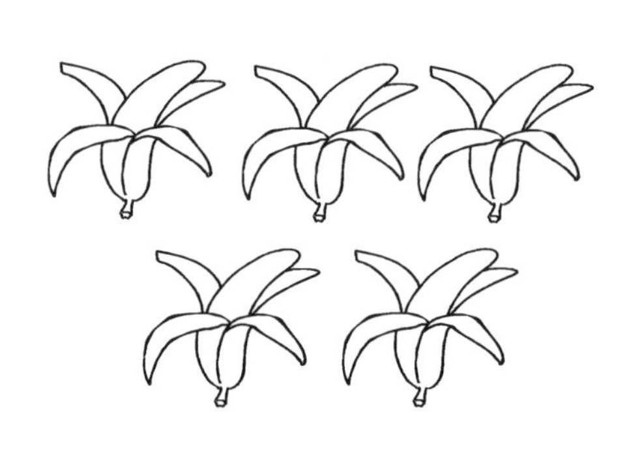

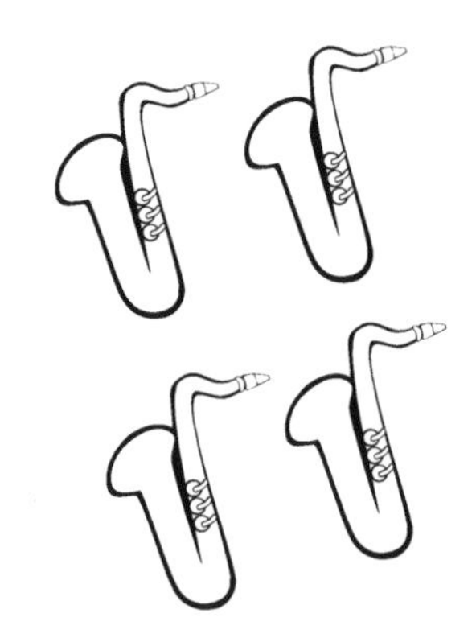

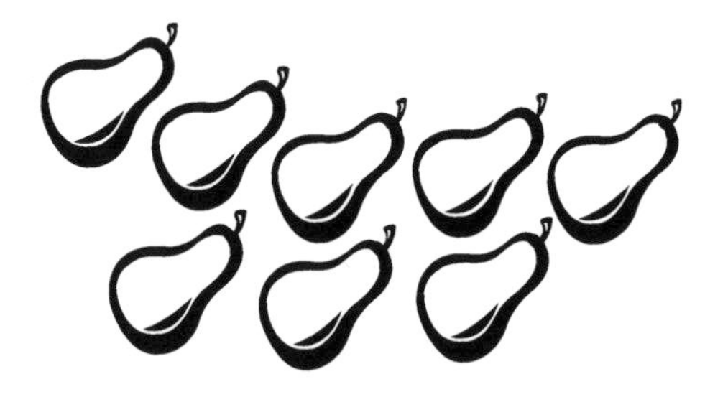

✷Complete the sets to make them eight.

8 trees

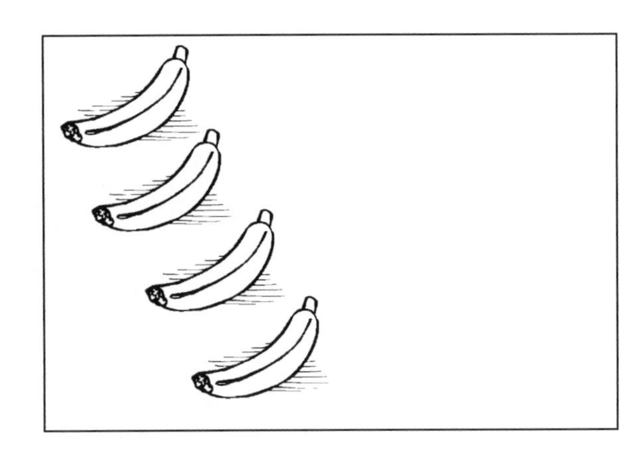

8 bananas

8 pigs

8 stars

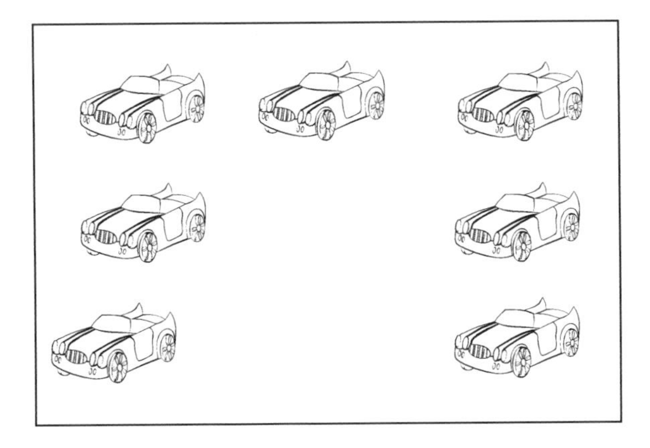

8 cars

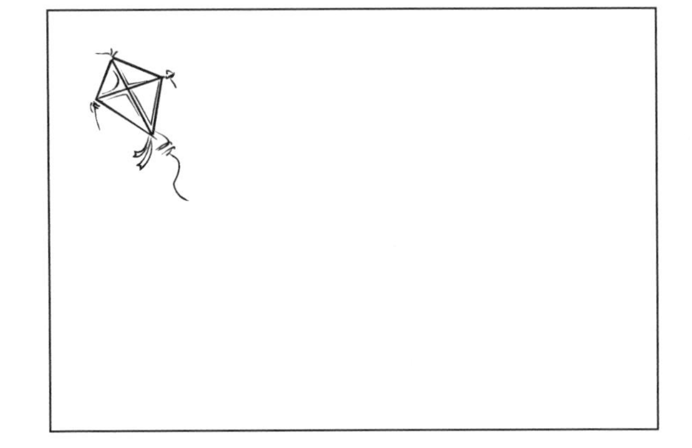

8 kites

✷ Count the objects. Read the number name and then colour the pictures.

✷Trace the nines.

9 9 9 9 9 9 9 9 9 9

9 9 9 9 9 9 9 9 9 9

✷Circle nine objects in each set.

✱Write some nines on the line. Start at the dot.

9

✱Draw objects in the boxes for the number.

9 crayons

9 pears

9 cups

9 eggs

9 balls

9 triangles △

✵Circle and then colour the sets that are nine.

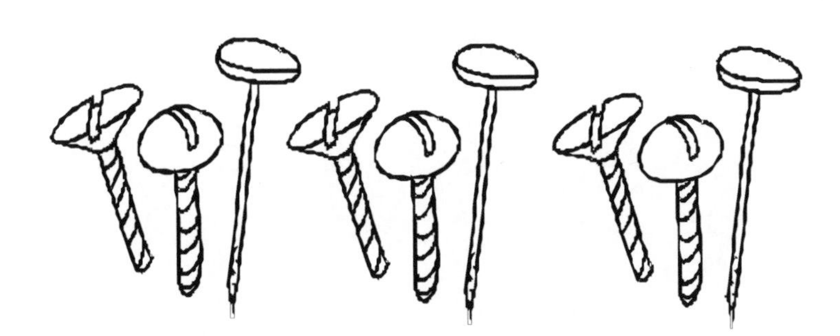

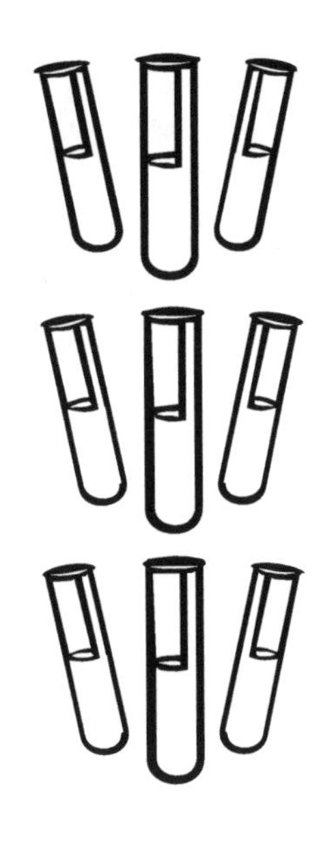

✱Complete the sets to make them nine.

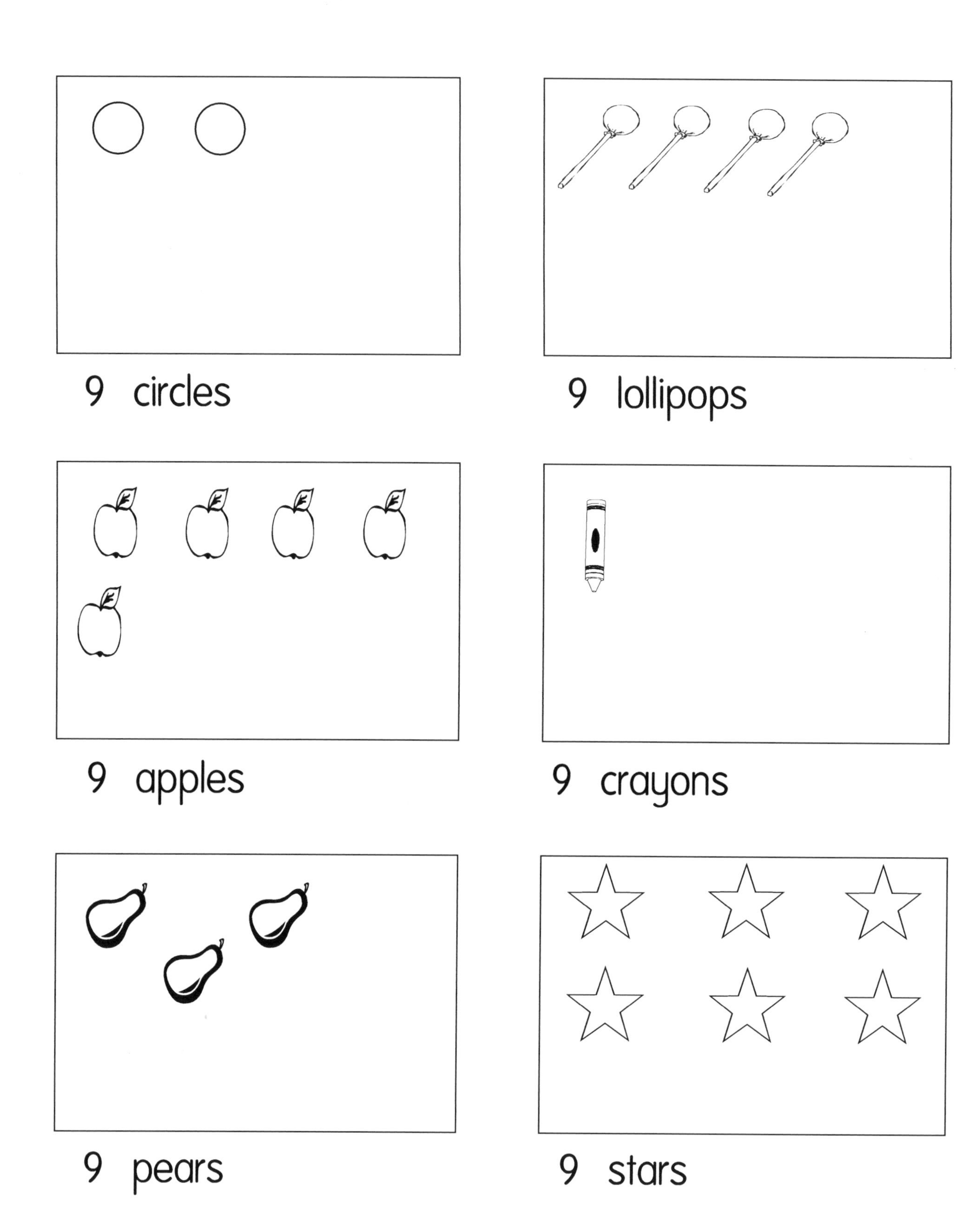

✸Count the objects. Read the number name and then colour the pictures.

10 ten

✵Trace the tens.

10 10 10 10 10 10 10 10

10 10 10 10 10 10 10 10

✵Circle ten objects in each set.

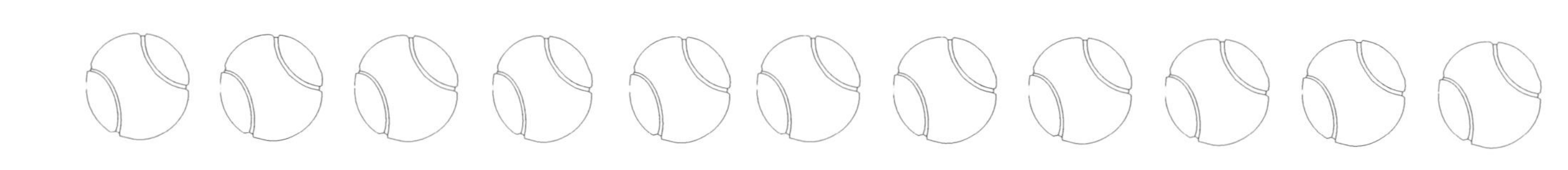

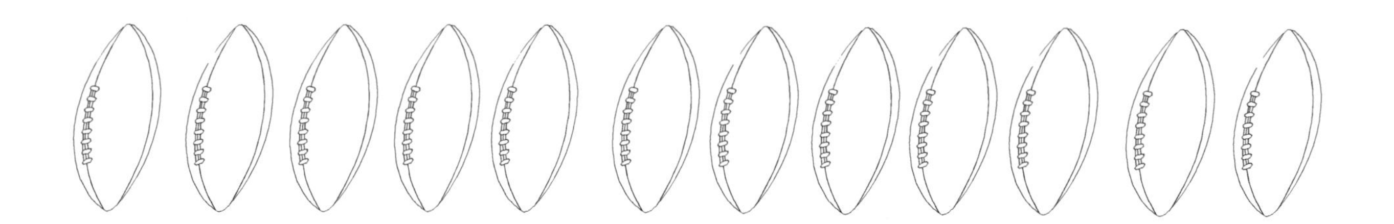

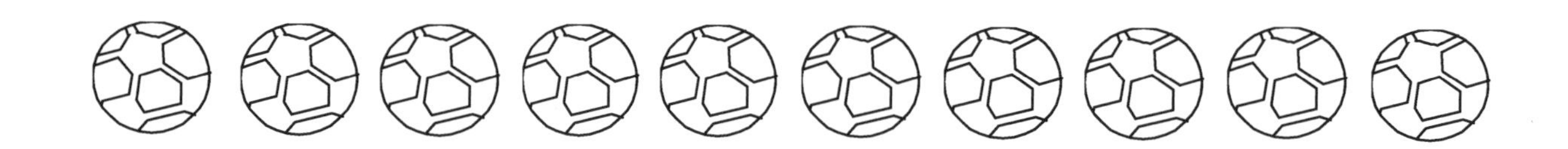

✱Write some tens on the line. Start at the dot.

10

✱Draw objects in the boxes for the number.

10 balls

10 eggs

10 apples

10 lollipops

10 pears

10 pencils 

✱Circle and then colour the sets that are ten.

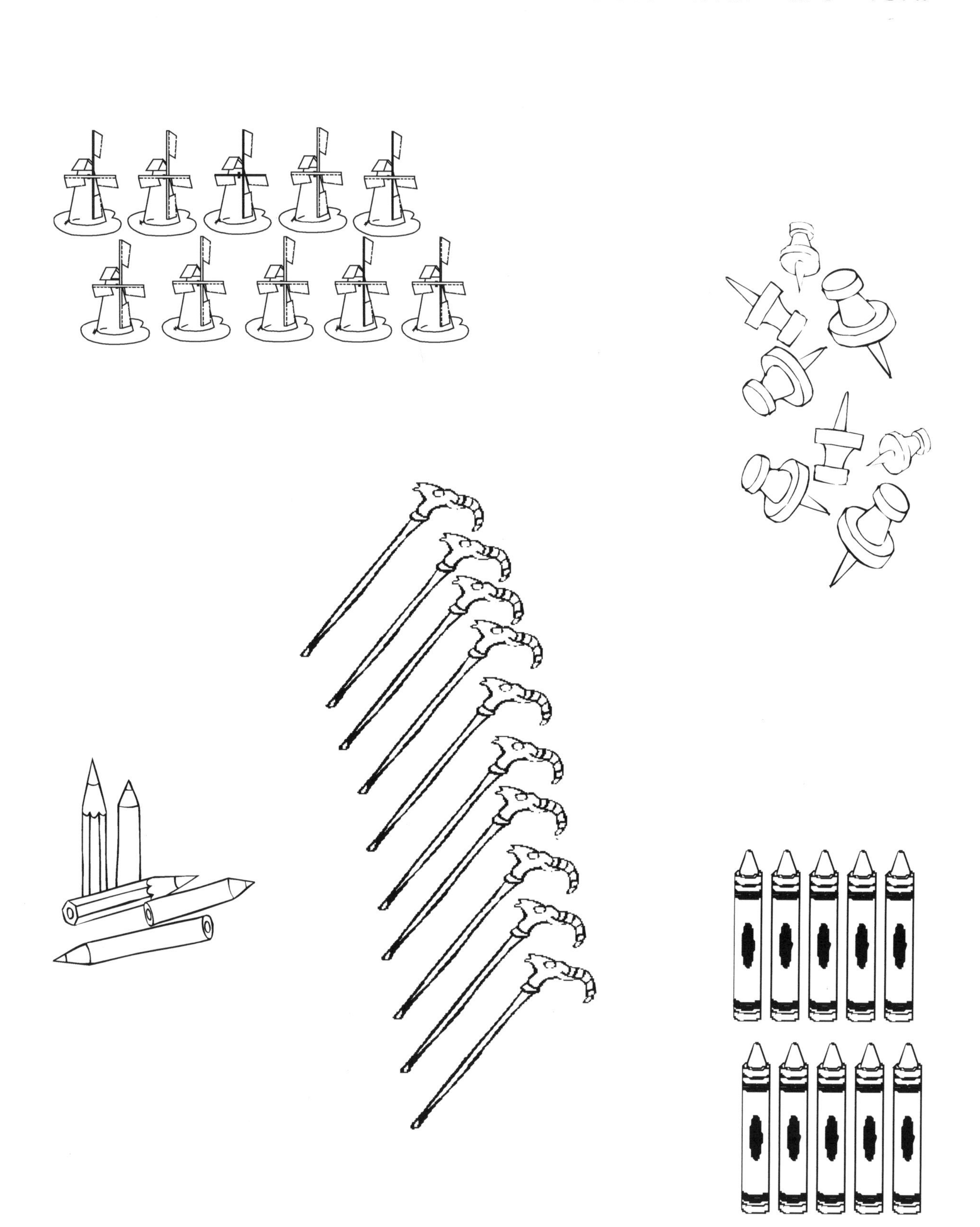

✷Complete the sets to make them ten.

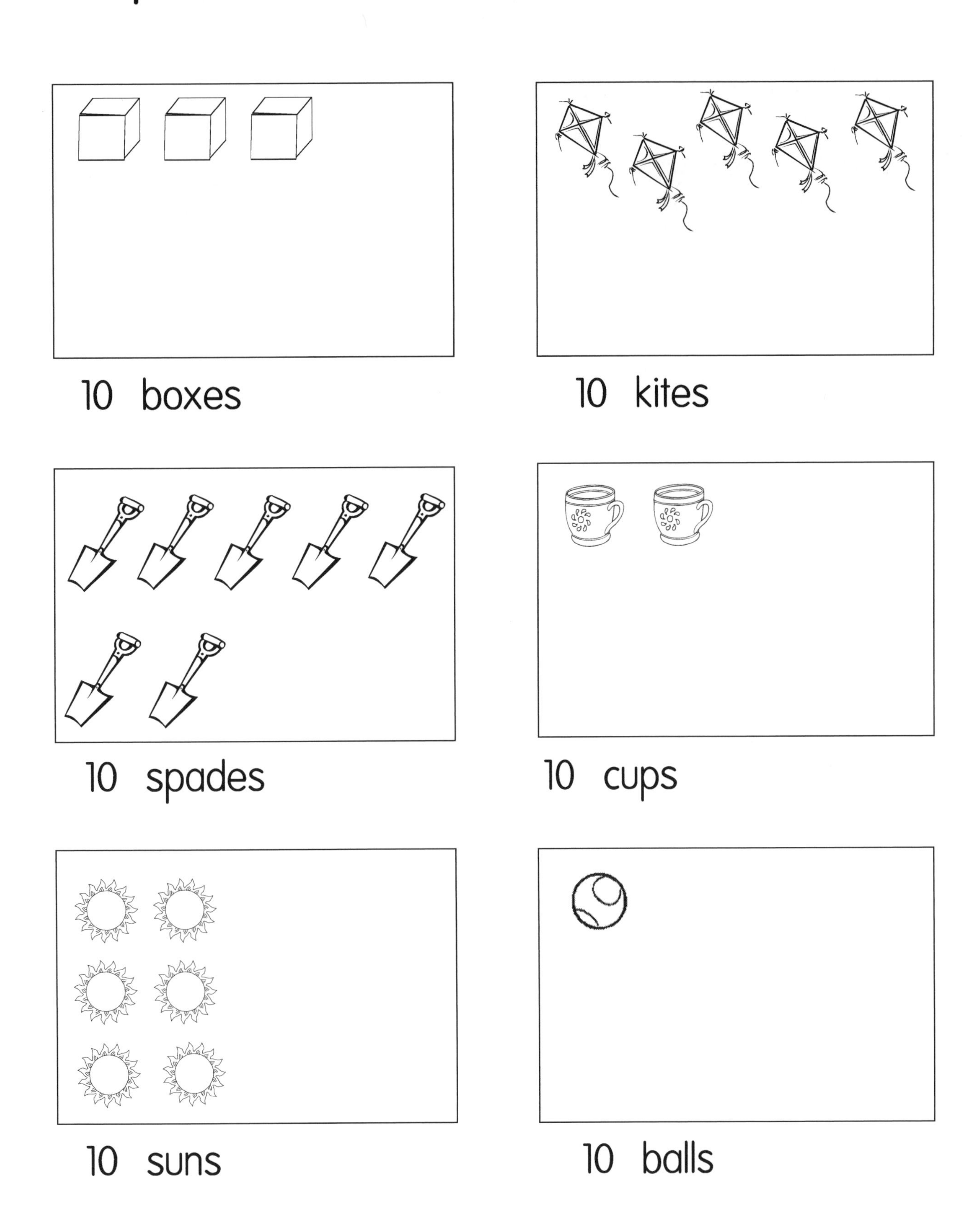

✱Circle the correct amount.

a) Circle six (6) cups.

b) Circle eight (8) glasses.

c) Circle ten (10) bears.

d) Circle nine (9) spades.

e) Circle seven (7) pigs.

✱Draw the objects for the numbers.

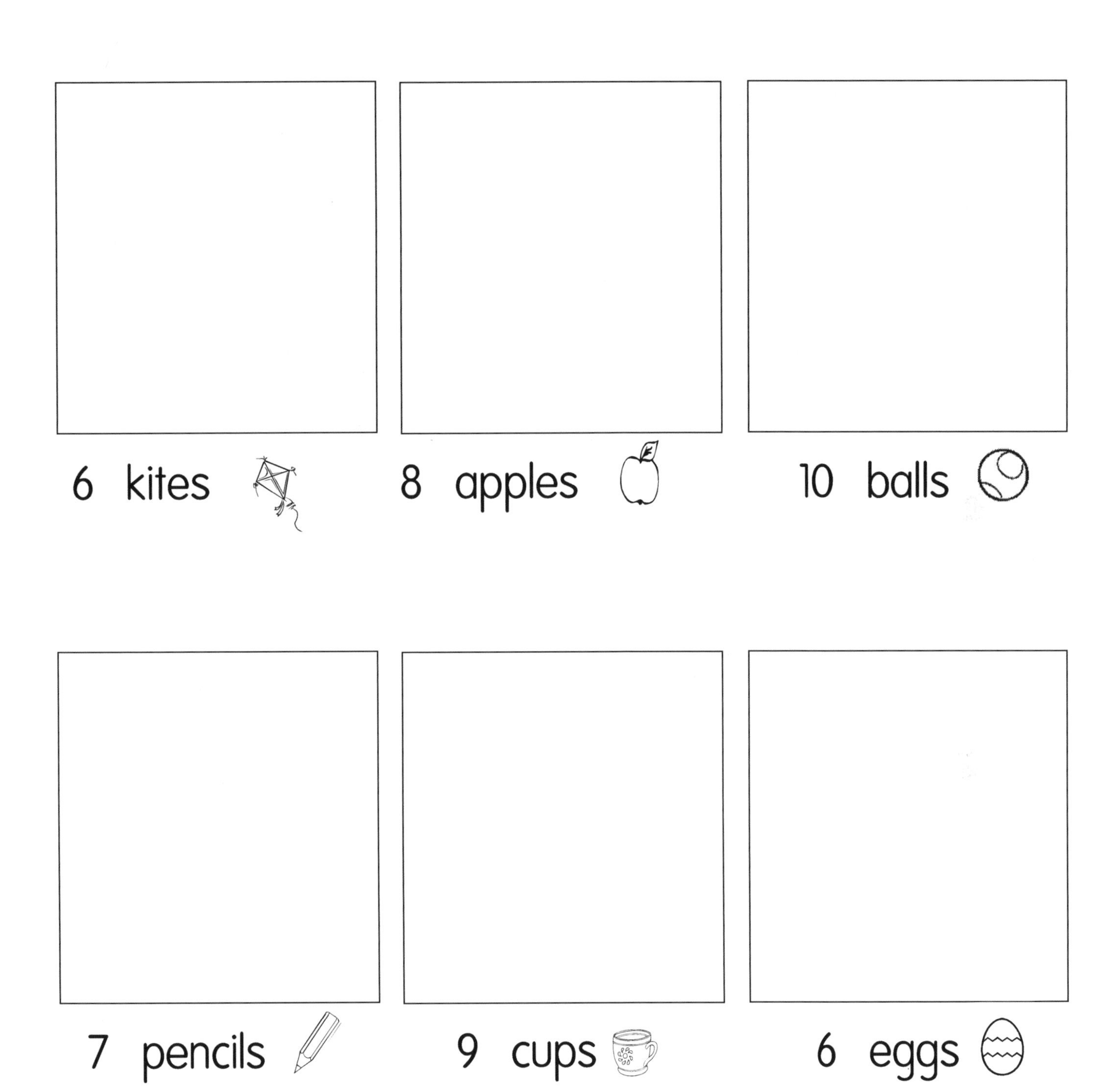

✶Complete the sets.

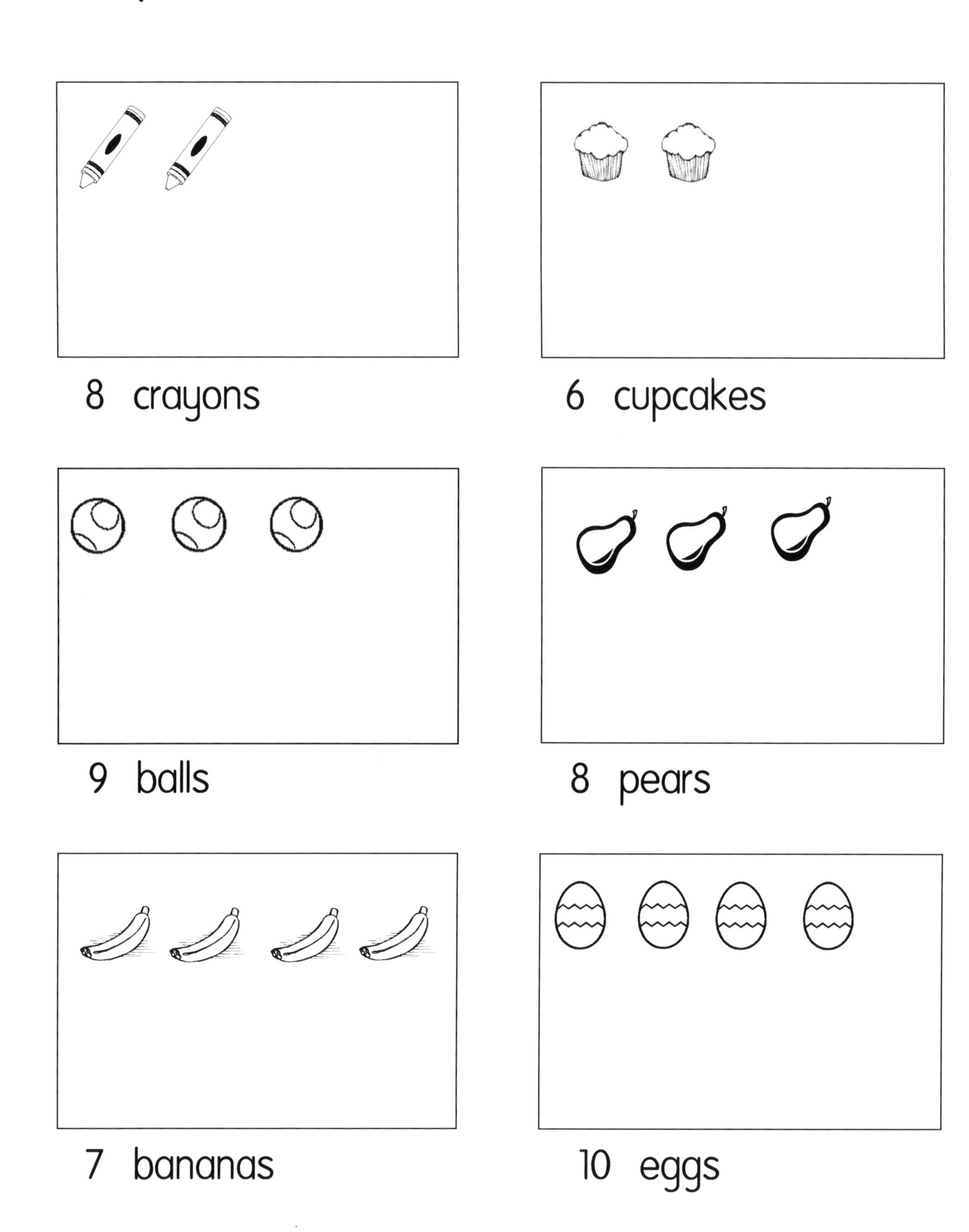

8 crayons

6 cupcakes

9 balls

8 pears

7 bananas

10 eggs

✱Count the objects and match them to the correct number.

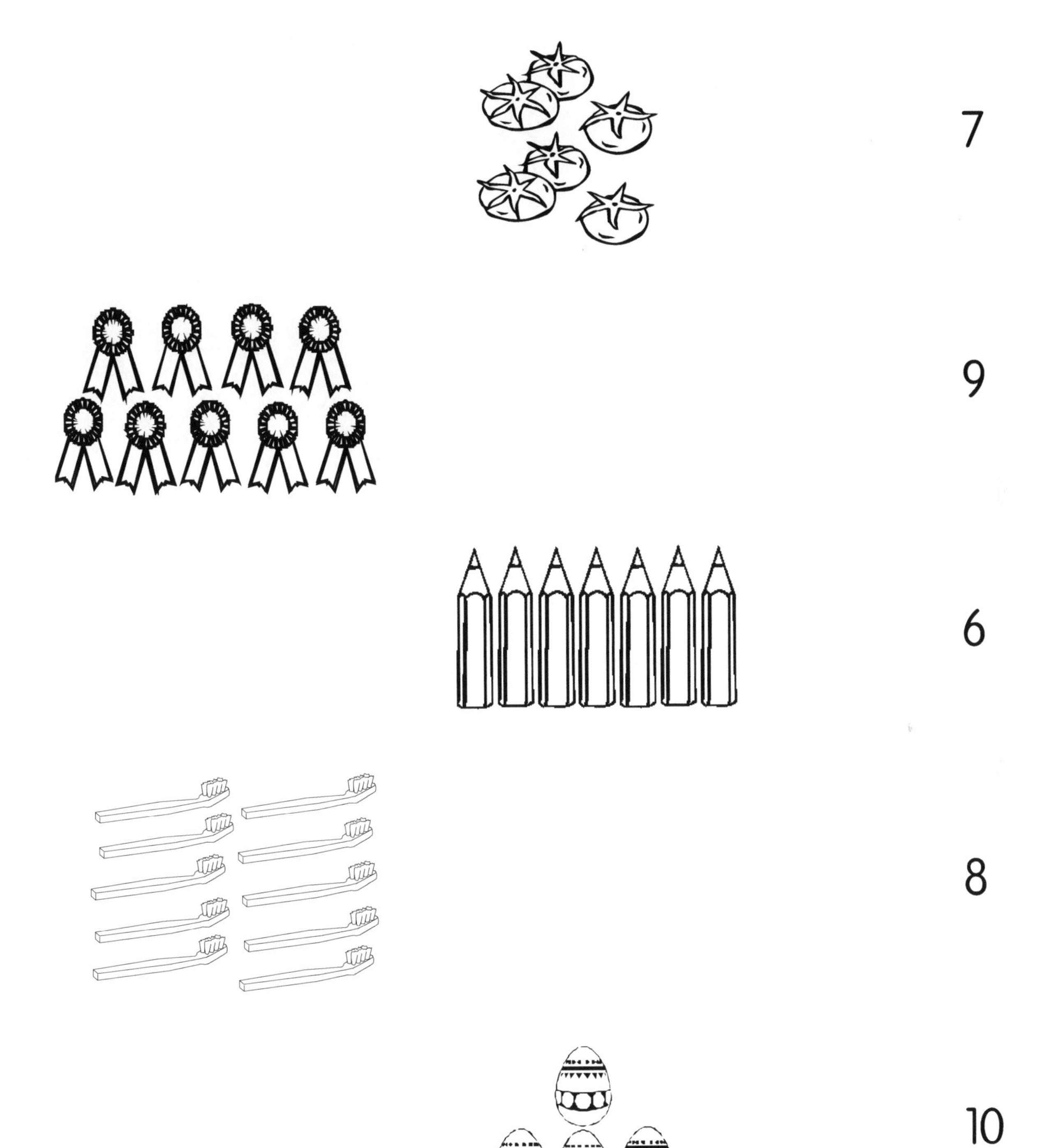

✱Count the objects then circle the correct number in the box.

5	4	8

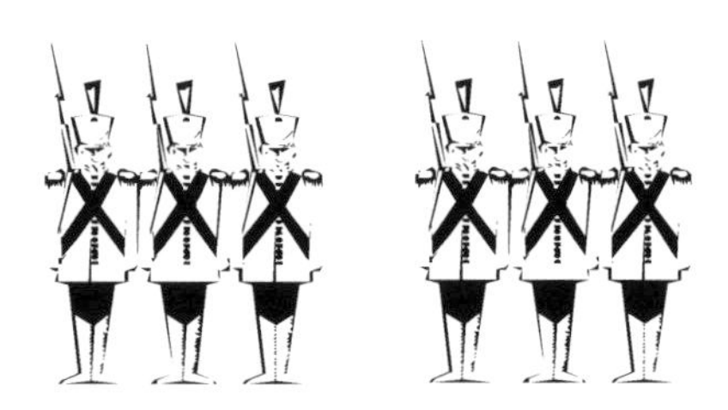

9	6	10

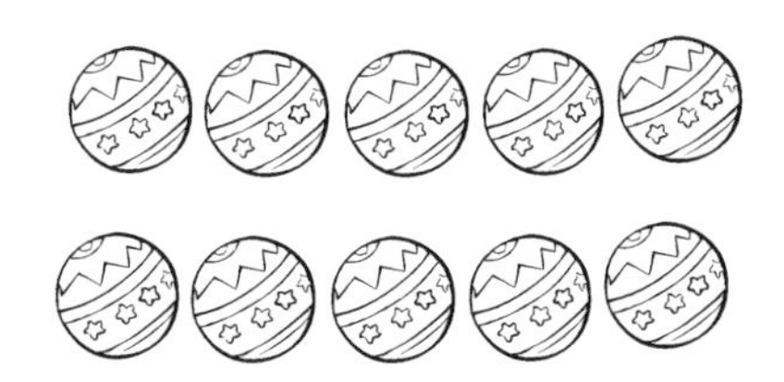

7	8	10

6	9	5

7	6	9

Part 3

✵Match the shapes that are the same.

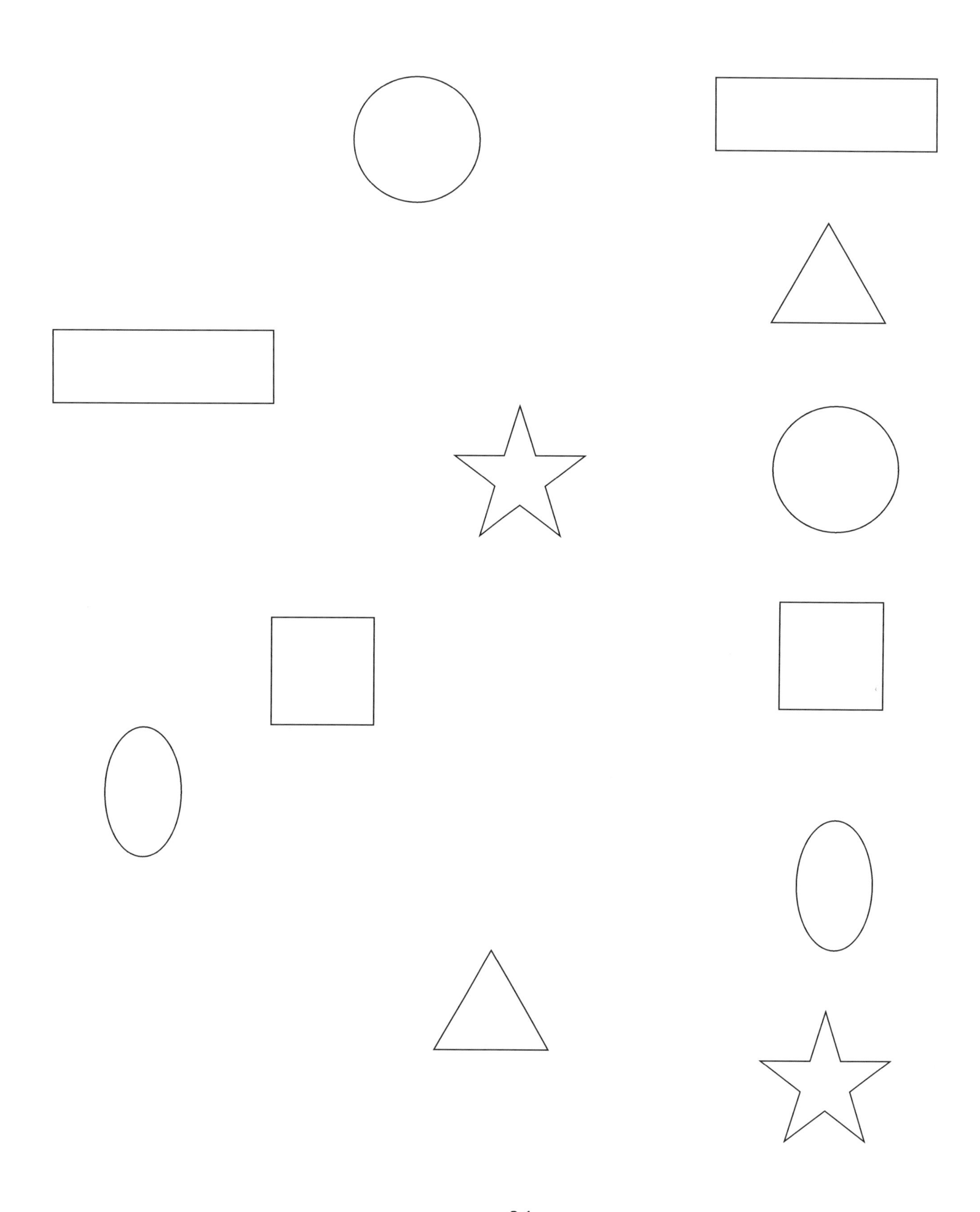

✱Trace the shapes.

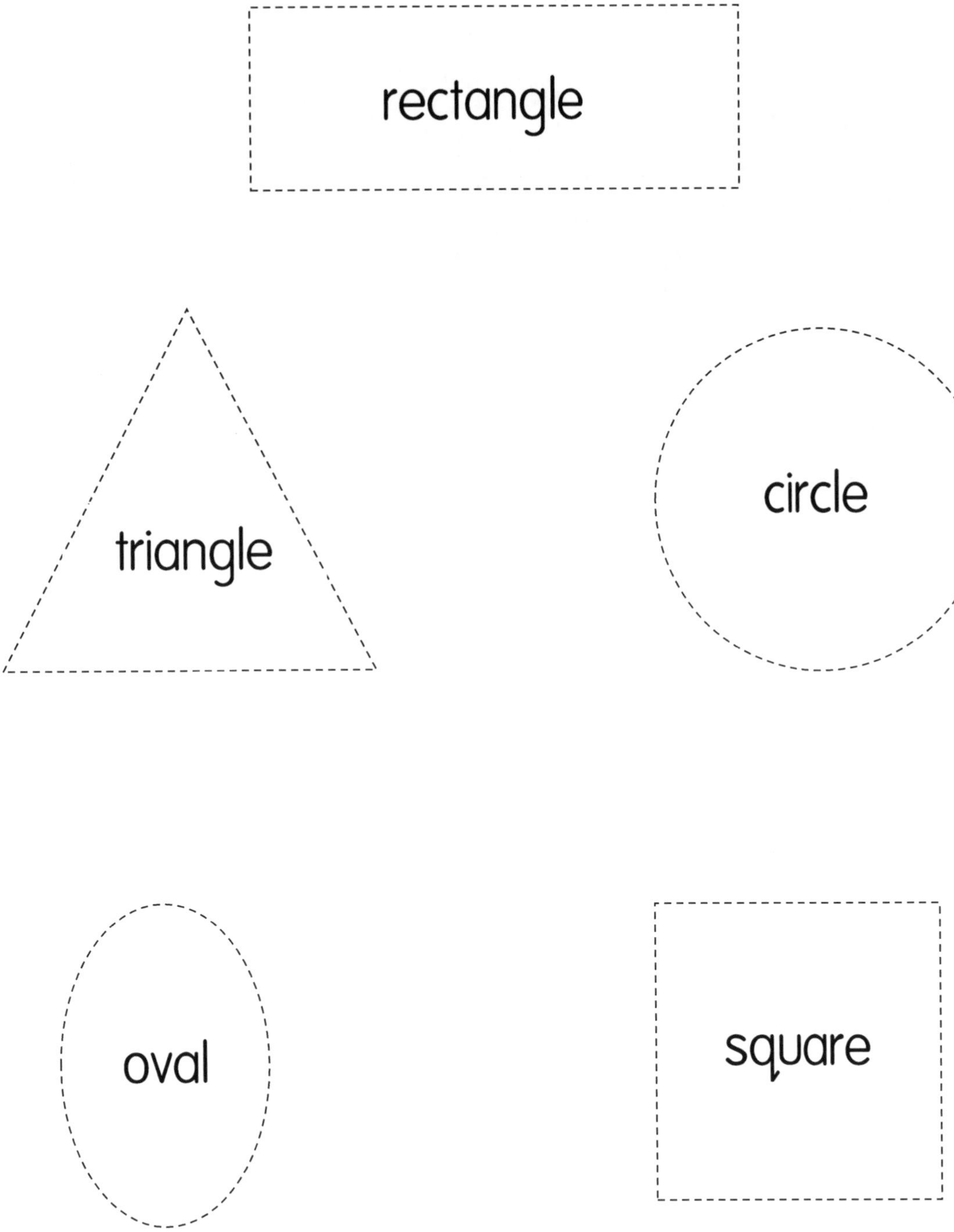

rectangle

triangle

circle

oval

square

✵ Trace each shape. Then draw some more on the line.

1. ______________________________

2. ______________________________

3. ______________________________

4. ______________________________

5. ______________________________

✻ Match each child to an ice cream.

✻ Match each bat to a ball.

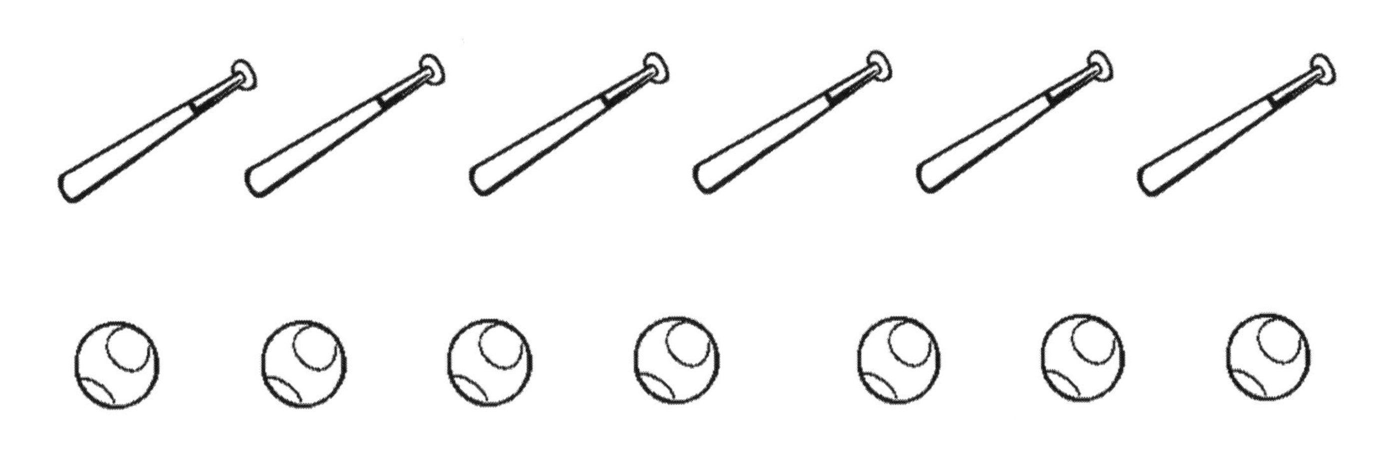

✻ Match each hammer to a nail.

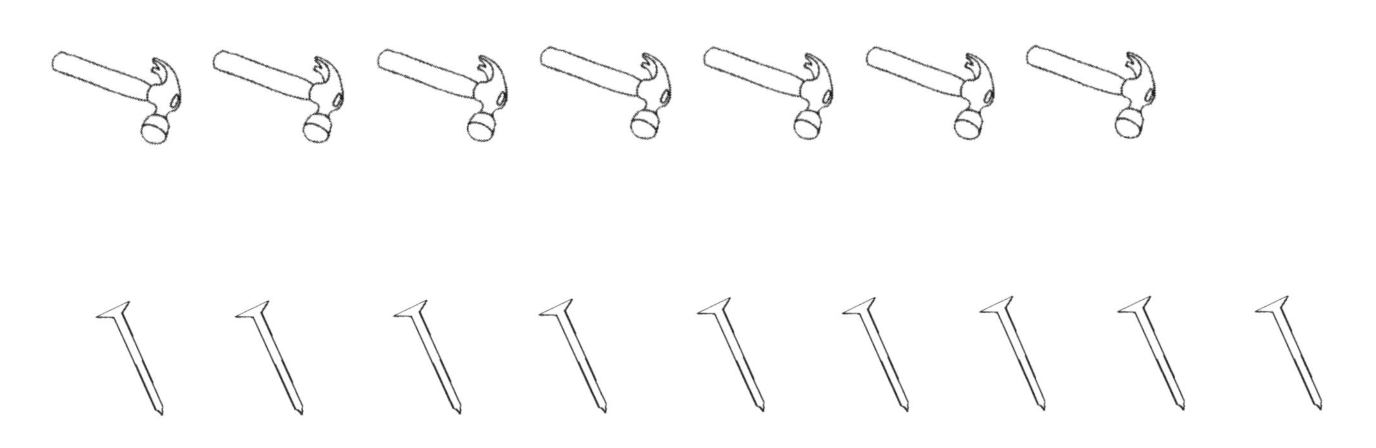

✴ Match the sets which have the same amount.

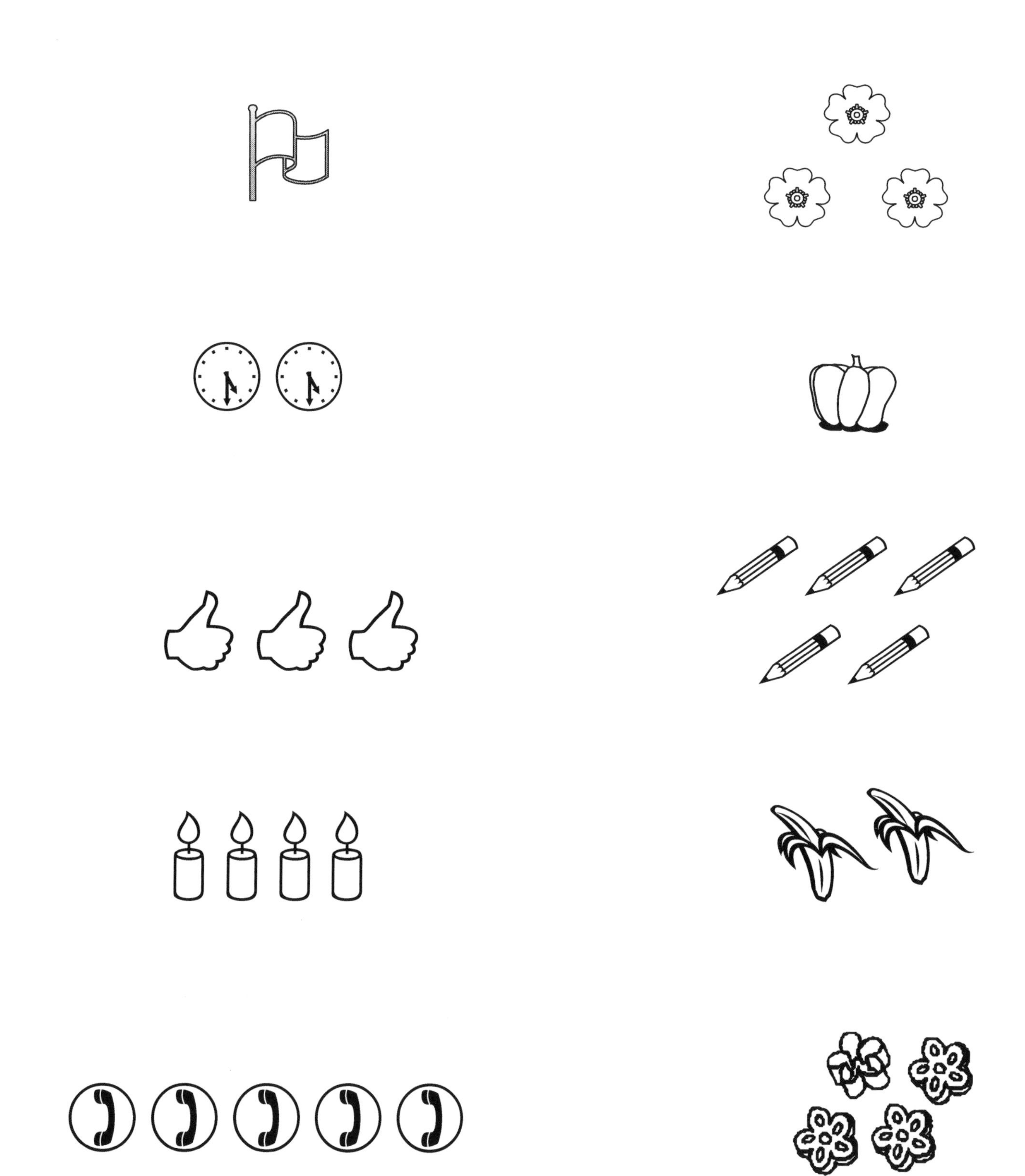

✱Make the bottom row the same as the top row.

a) 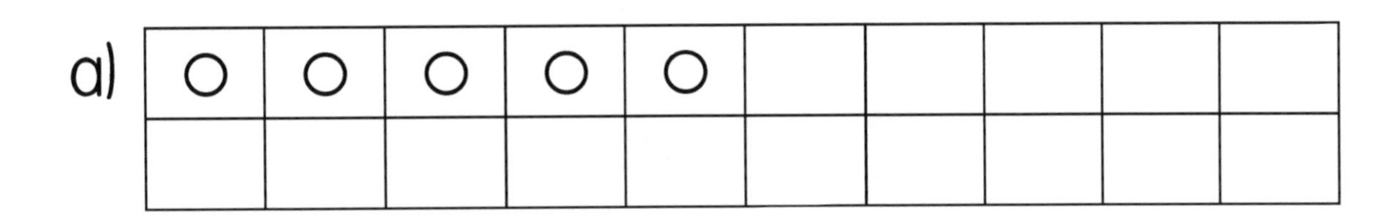

b) 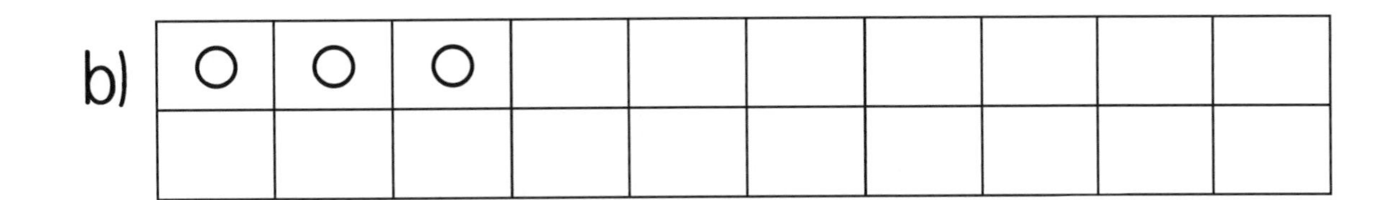

c)

○	○	○	○	○	○	○	○		

d)

○	○	○	○	○	○				

e)

○	○	○	○						

f)

○	○	○	○	○	○	○	○	○	○

✵ In the boxes, draw objects to make the sets equal

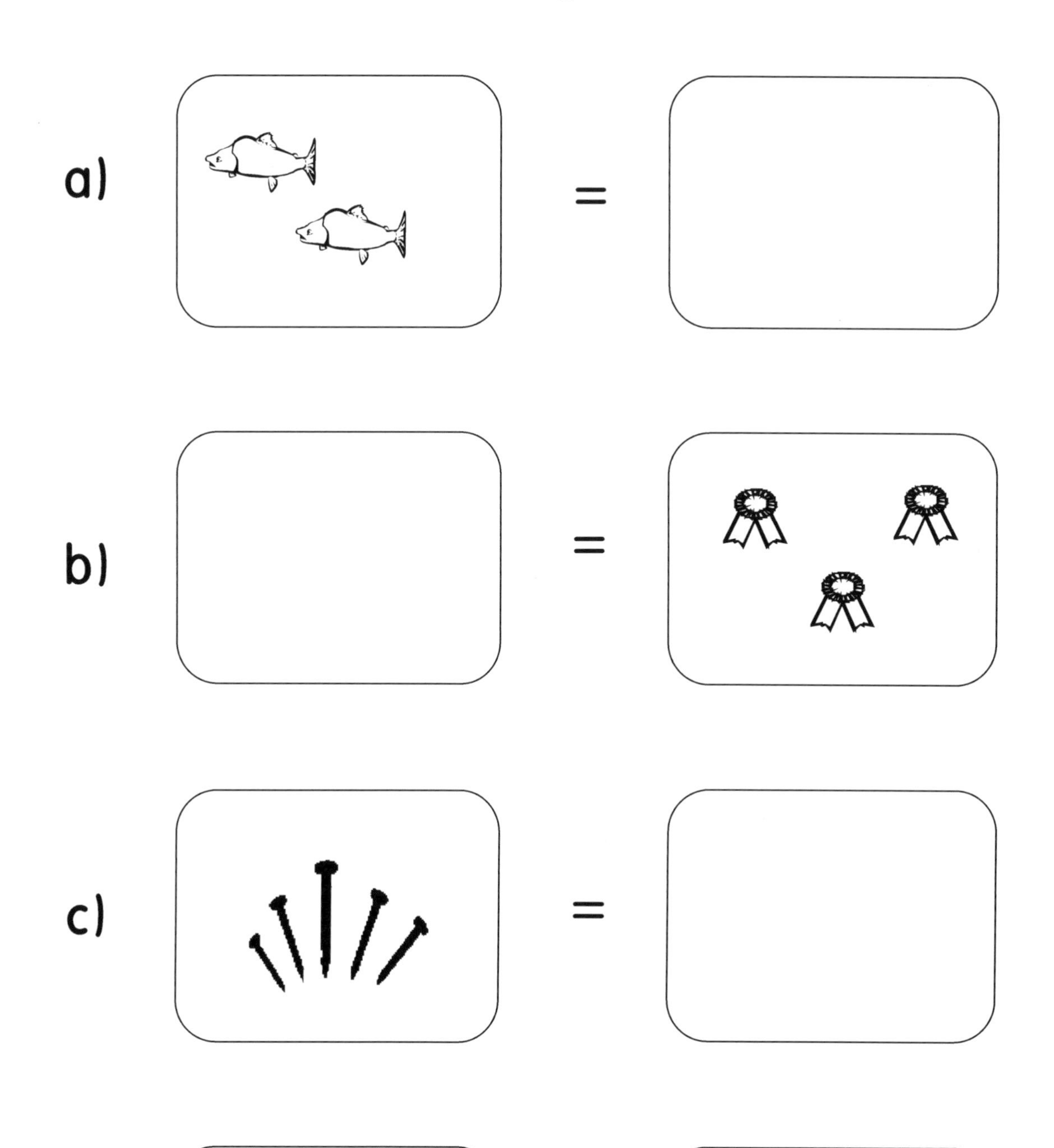

✵Make the sets equal. Then put in the sign.

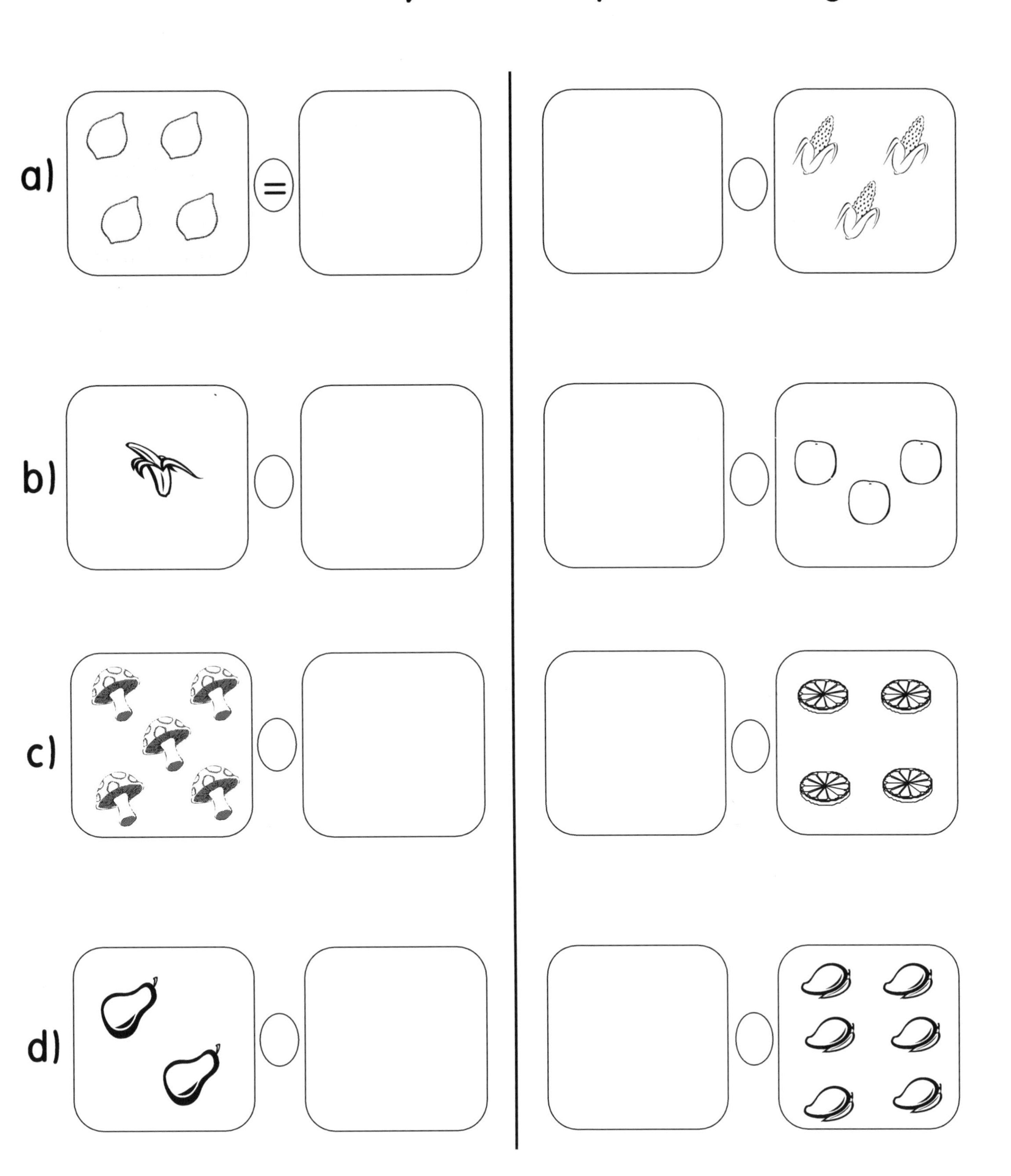

✱ Circle the sets which are more.

a)  

b)  

c) 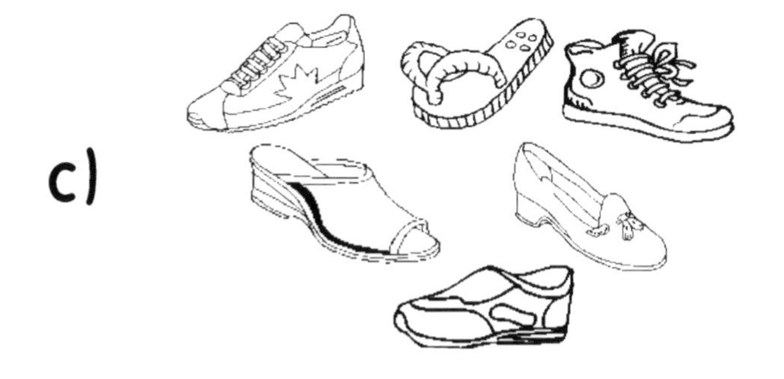 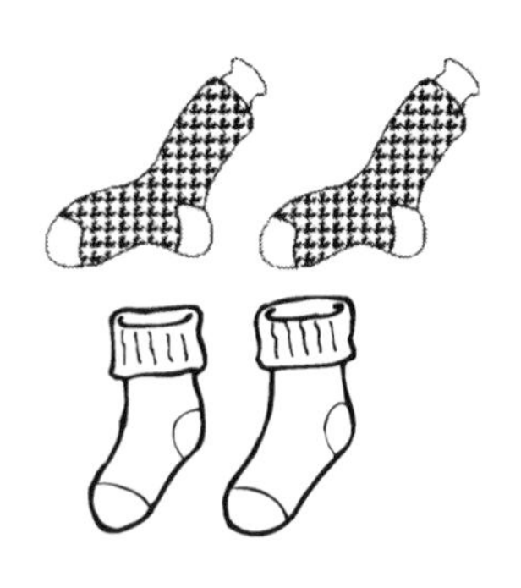

✱Make the bottom row more than the top row.

a)

O	O	O	O						

b)

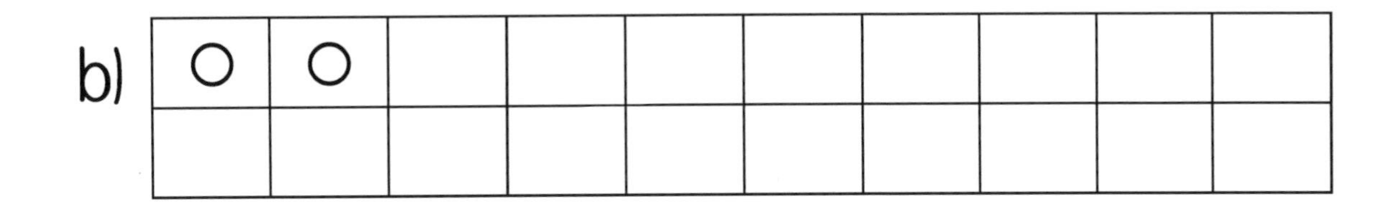

c)

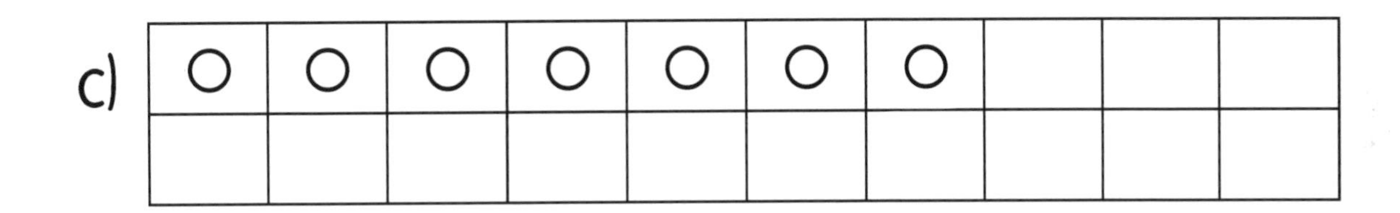

d)

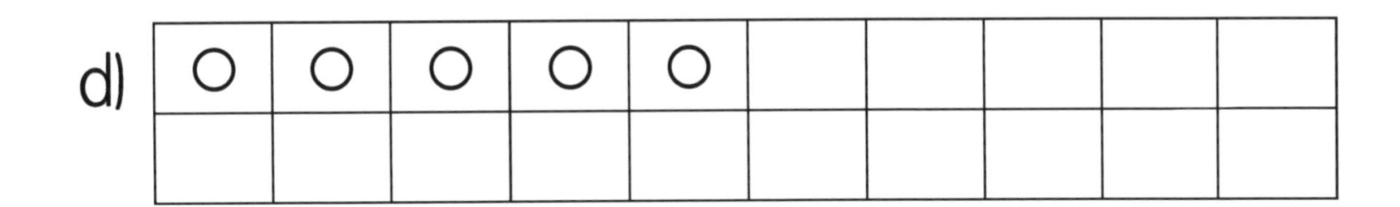

e)

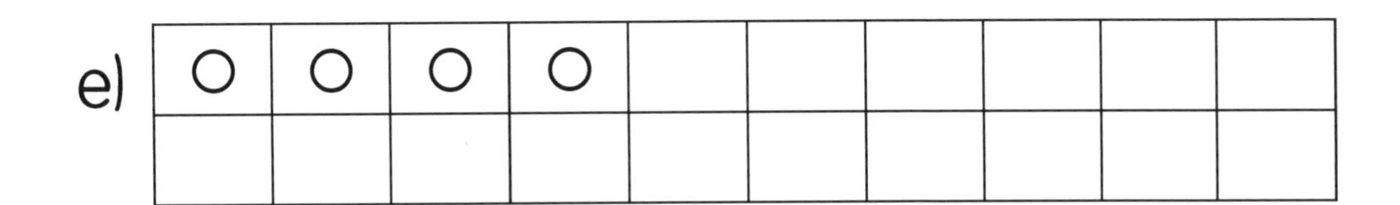

f)

O	O	O	O	O	O	O	O		

✱ In the boxes, draw objects to make the sets more.

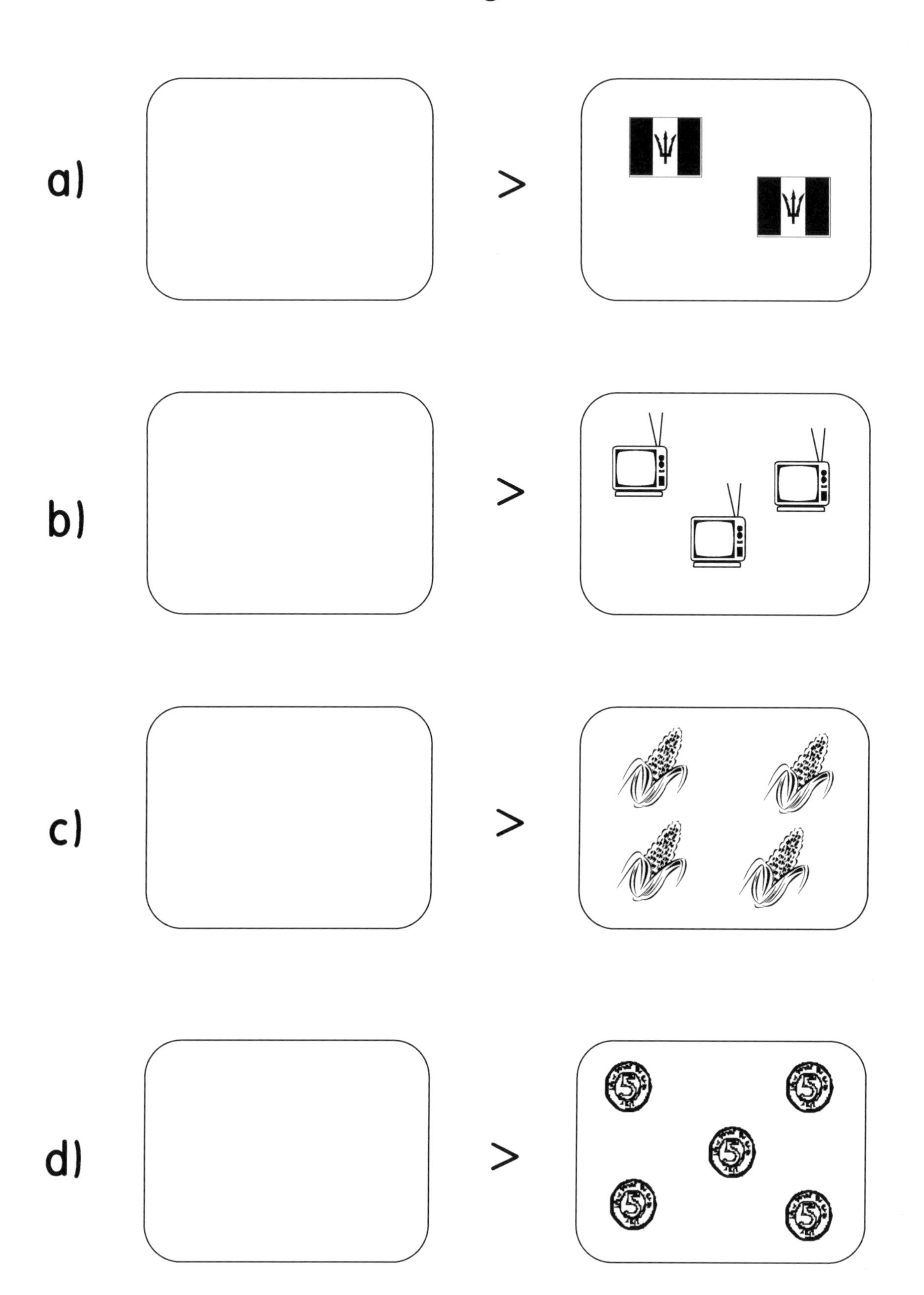

✱ Make the sets more. Then put in the sign.

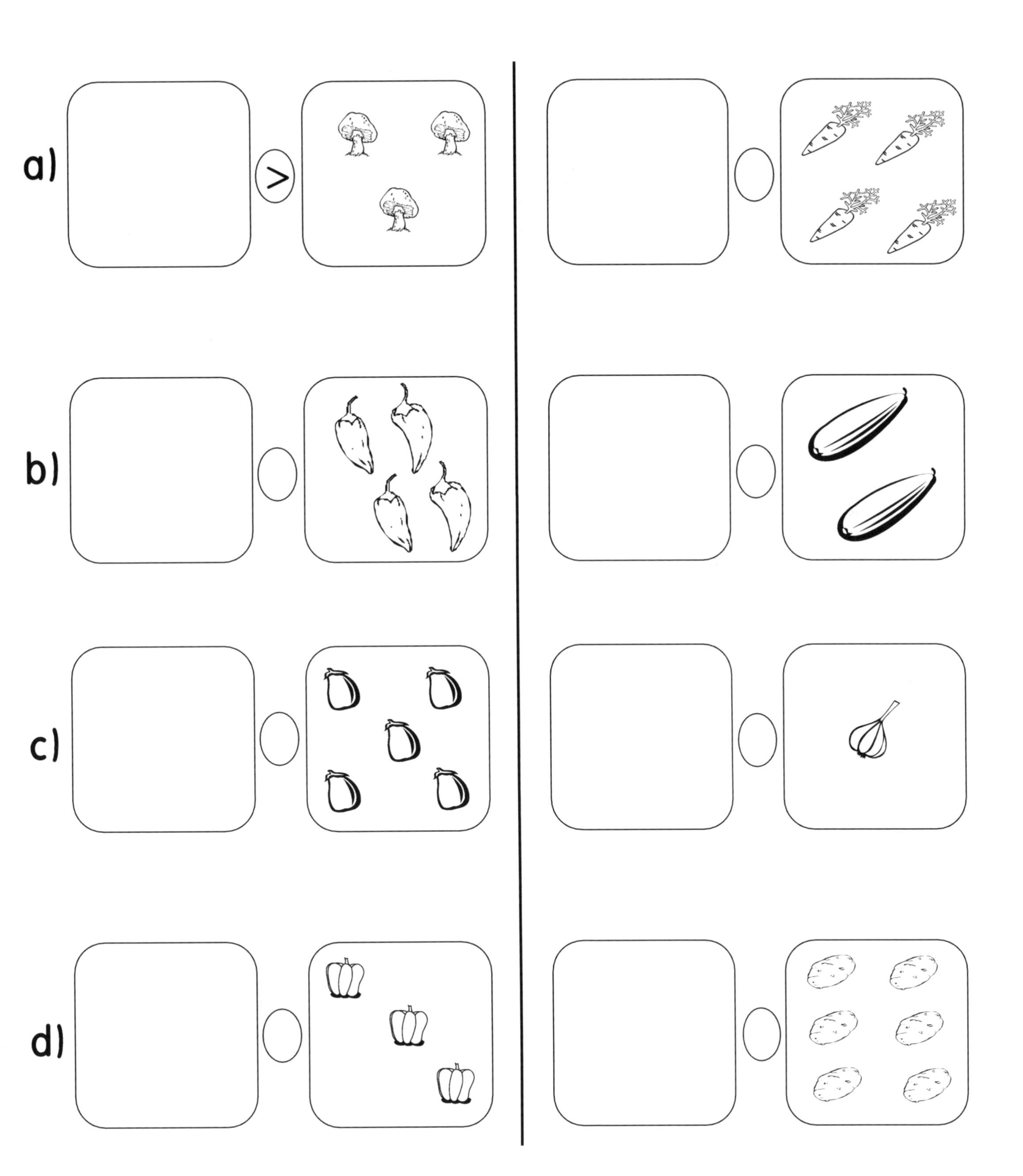

✵ Circle the sets which are less.

a) 

b) 

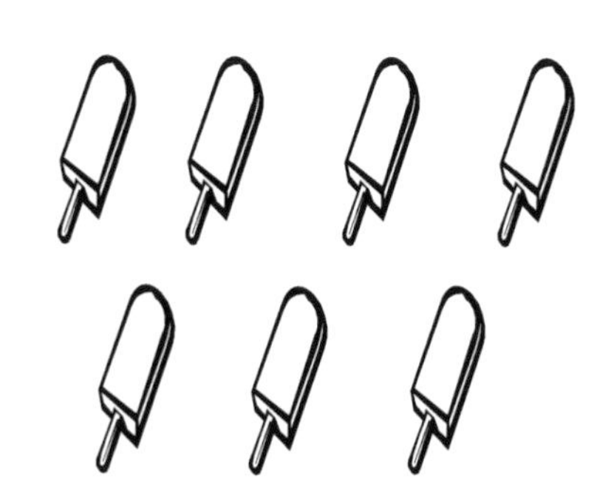

c) 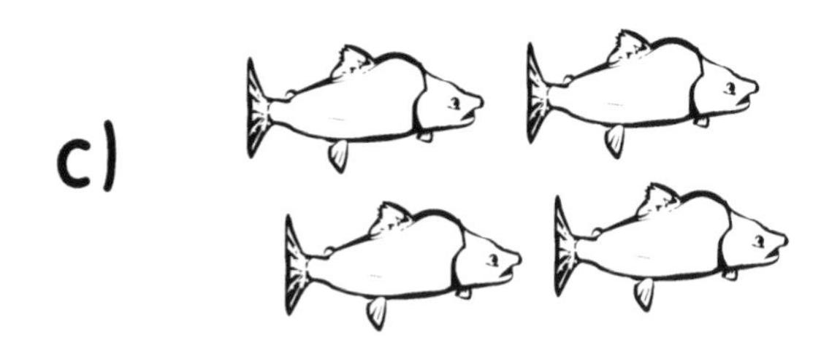

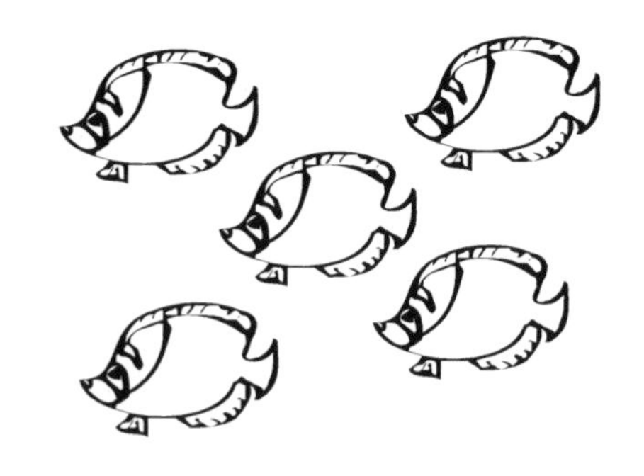

✷ Make the bottom row less than the top row.

a)

○	○	○	○	○	○				

b)

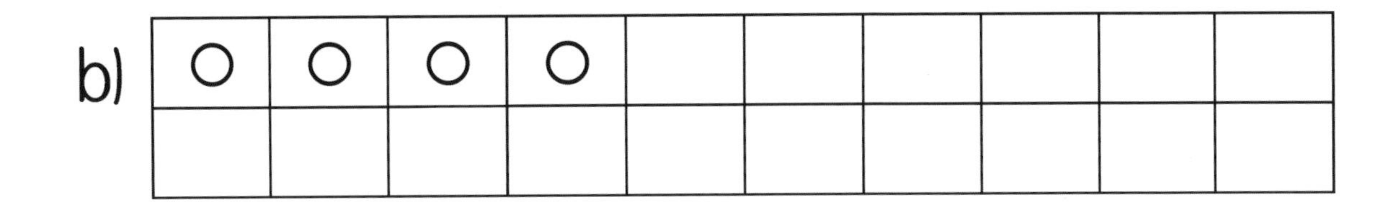

c)

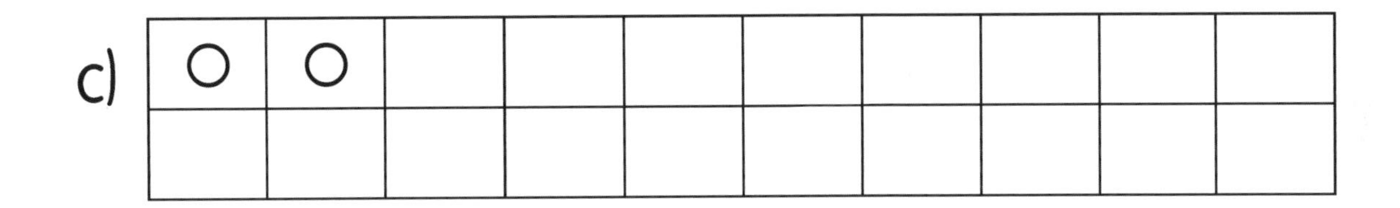

d)

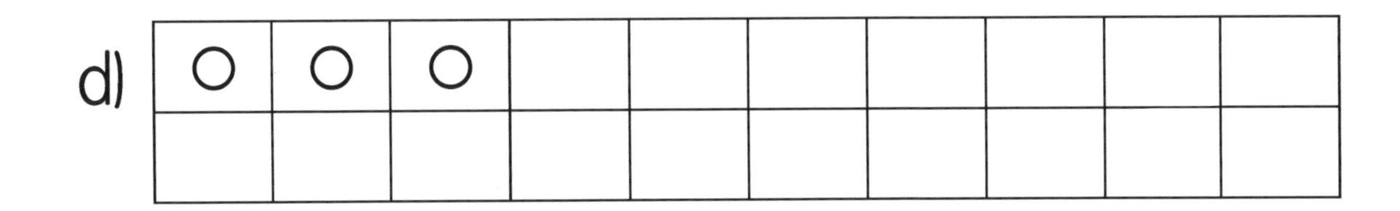

e)

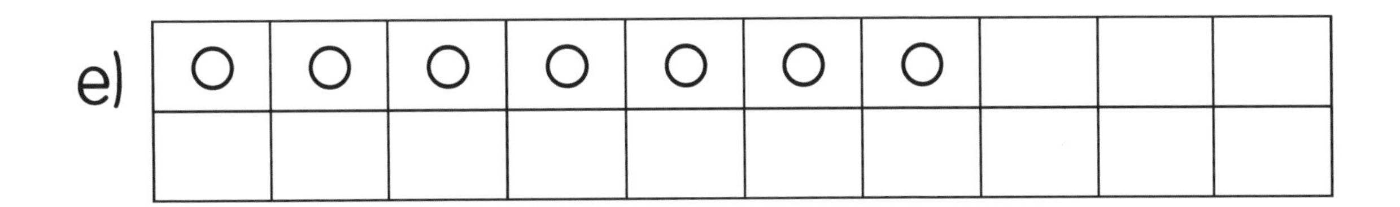

f)

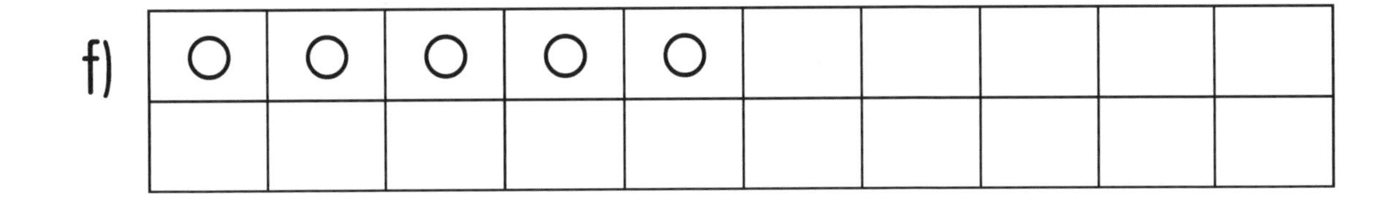

✵ In the boxes, draw objects to make the sets less.

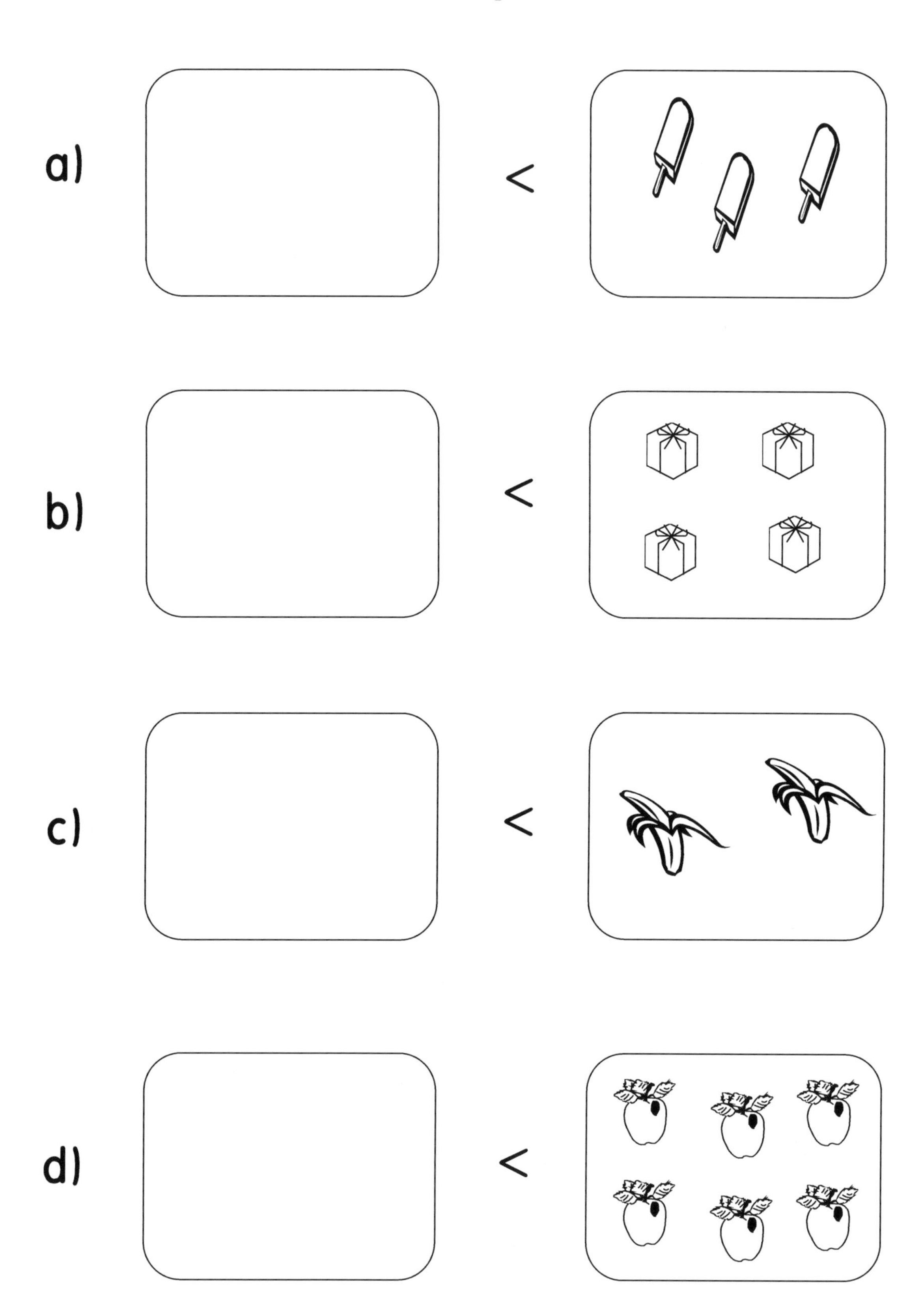

✷Make the sets less. Then put in the sign.

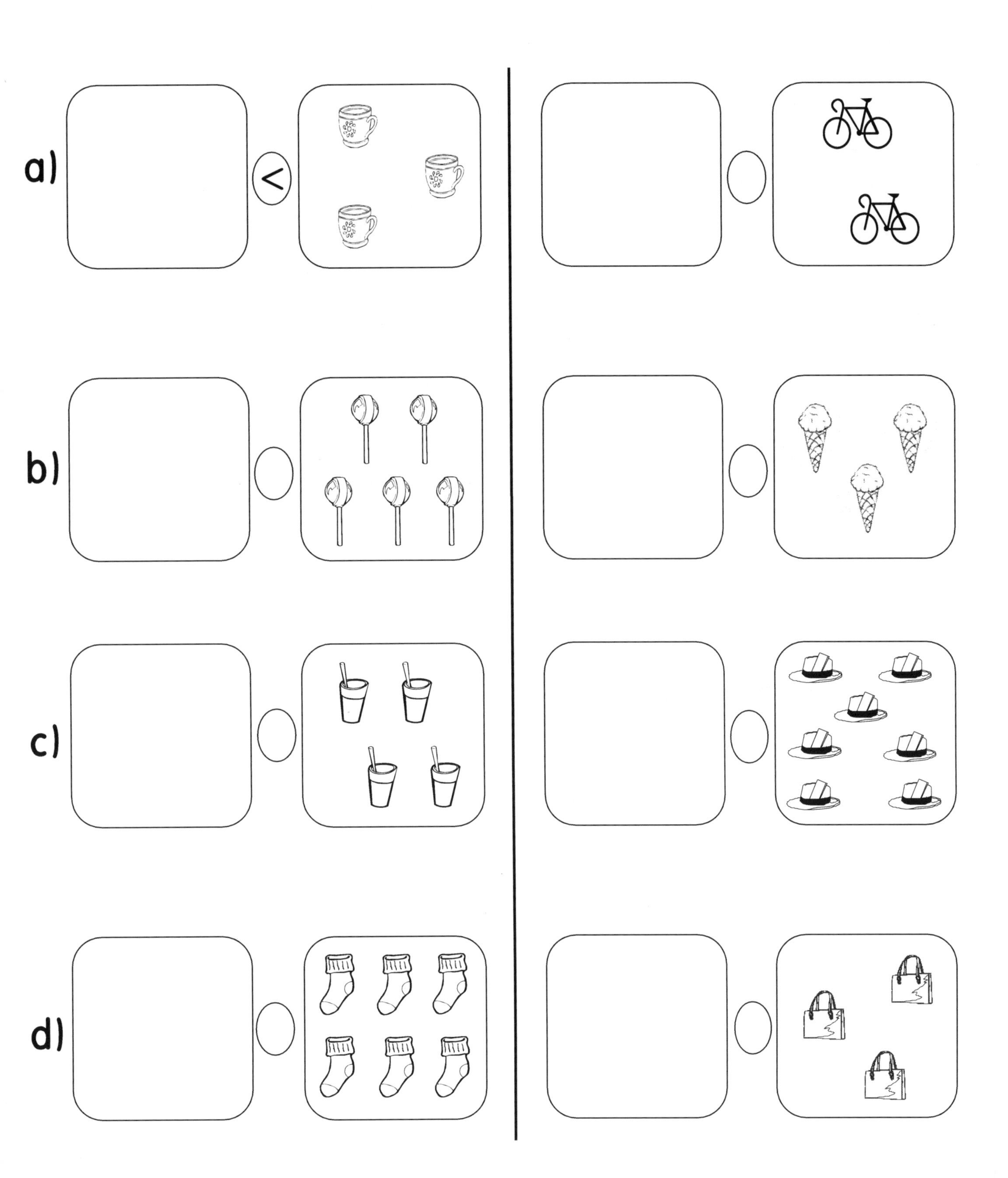

✱Put in the correct sign in the ovals. (< > =)

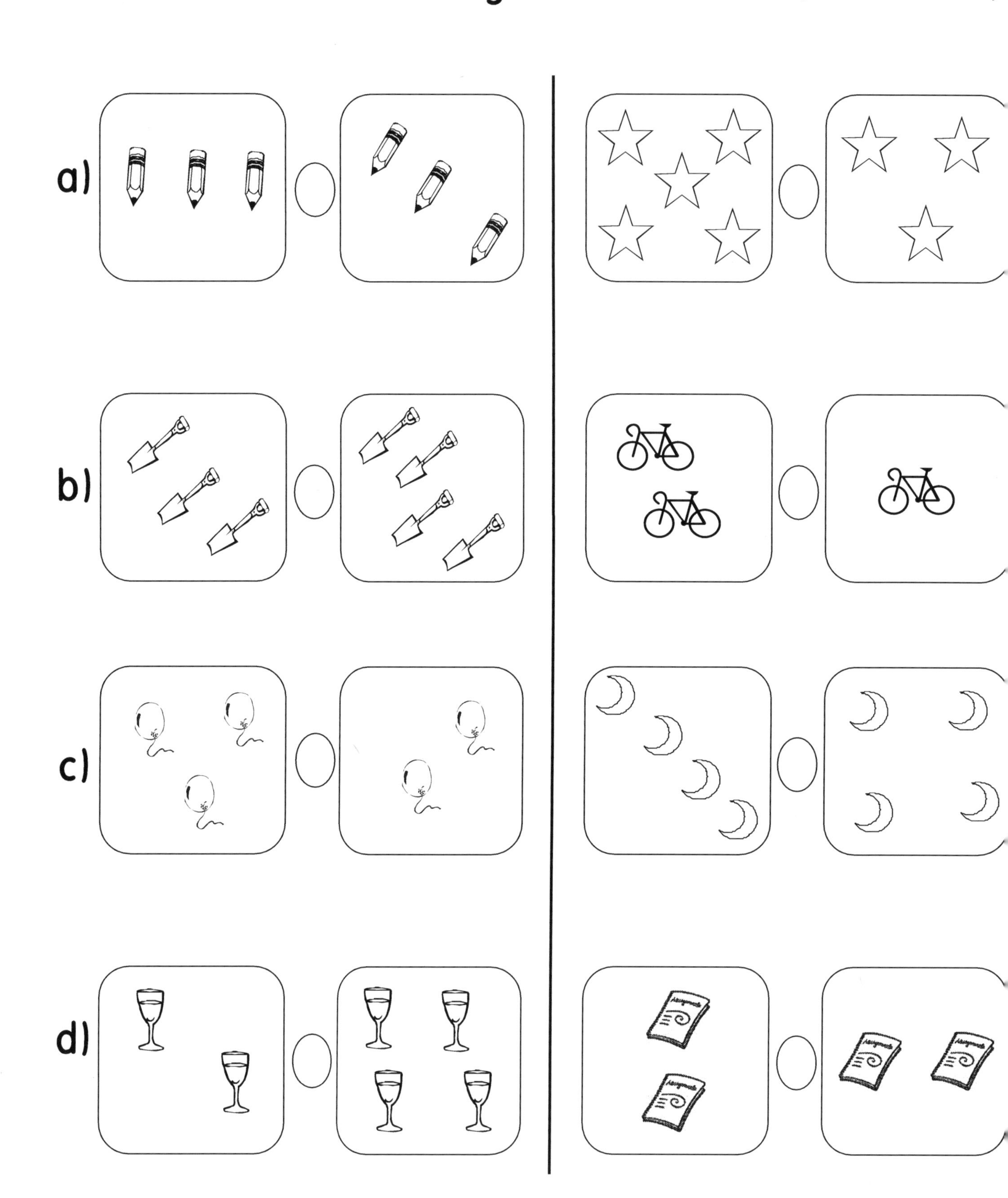

✱Circle the bigger number in each can.

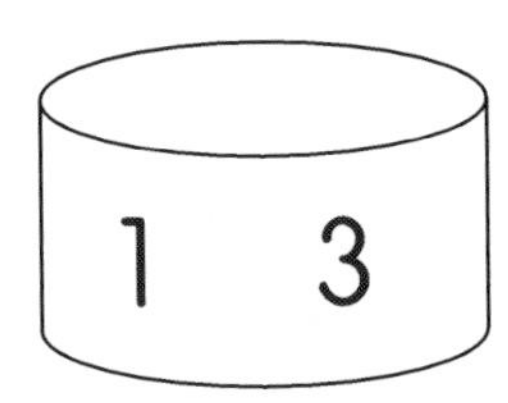

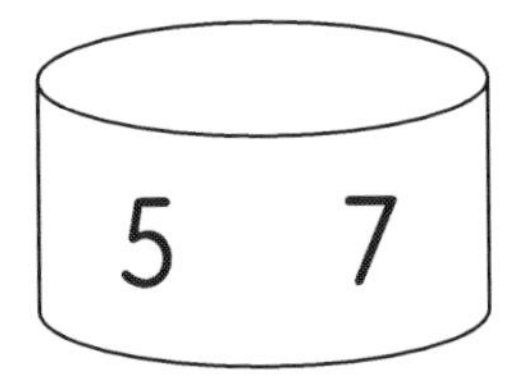

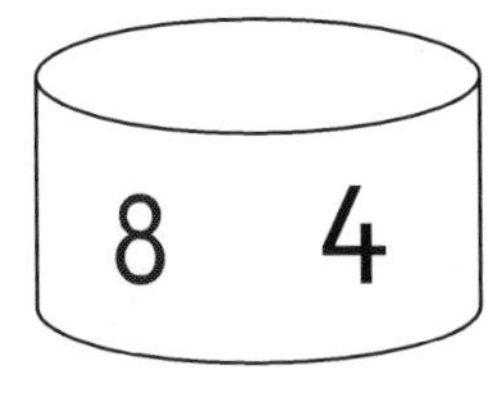

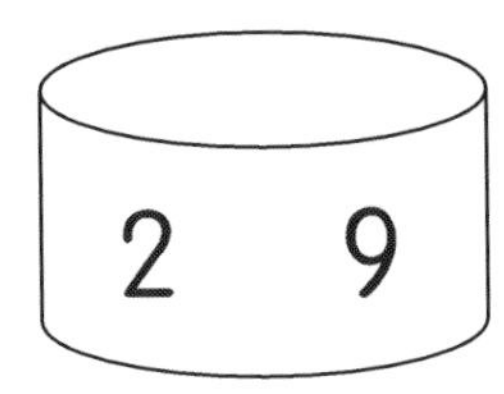

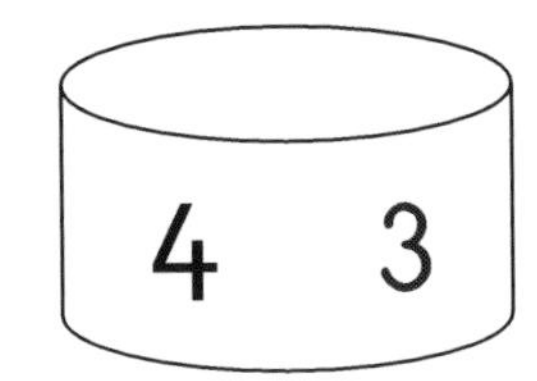

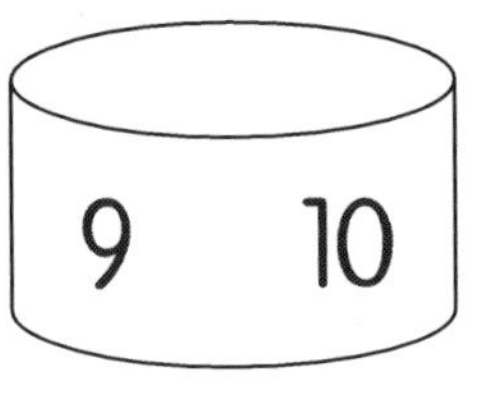

✱Circle the smaller number in each box.

3 2	5 6	0 1
6 8	9 7	4 3

✱Put the correct sign in the box. ($<$ $>$ $=$)

 ☐ 3

 ☐ 6

☐ 6

 ☐ 8

7 ☐

5 ☐

9 ☐

4 ☐

✱Put the correct sign in the box. ($<$ $>$ $=$)

$4 \ \square \ 3$ $\qquad$ $2 \ \square \ 5$

$7 \ \square \ 8$ $\qquad$ $3 \ \square \ 3$

$5 \ \square \ 5$ $\qquad$ $9 \ \square \ 6$

✱Write a suitable number in the box.

$\square < 5$ $\qquad$ $2 > \square$

$\square > 5$ $\qquad$ $2 = \square$

$\square = 5$ $\qquad$ $2 < \square$

✵Look at the signs and then draw objects in the boxes.

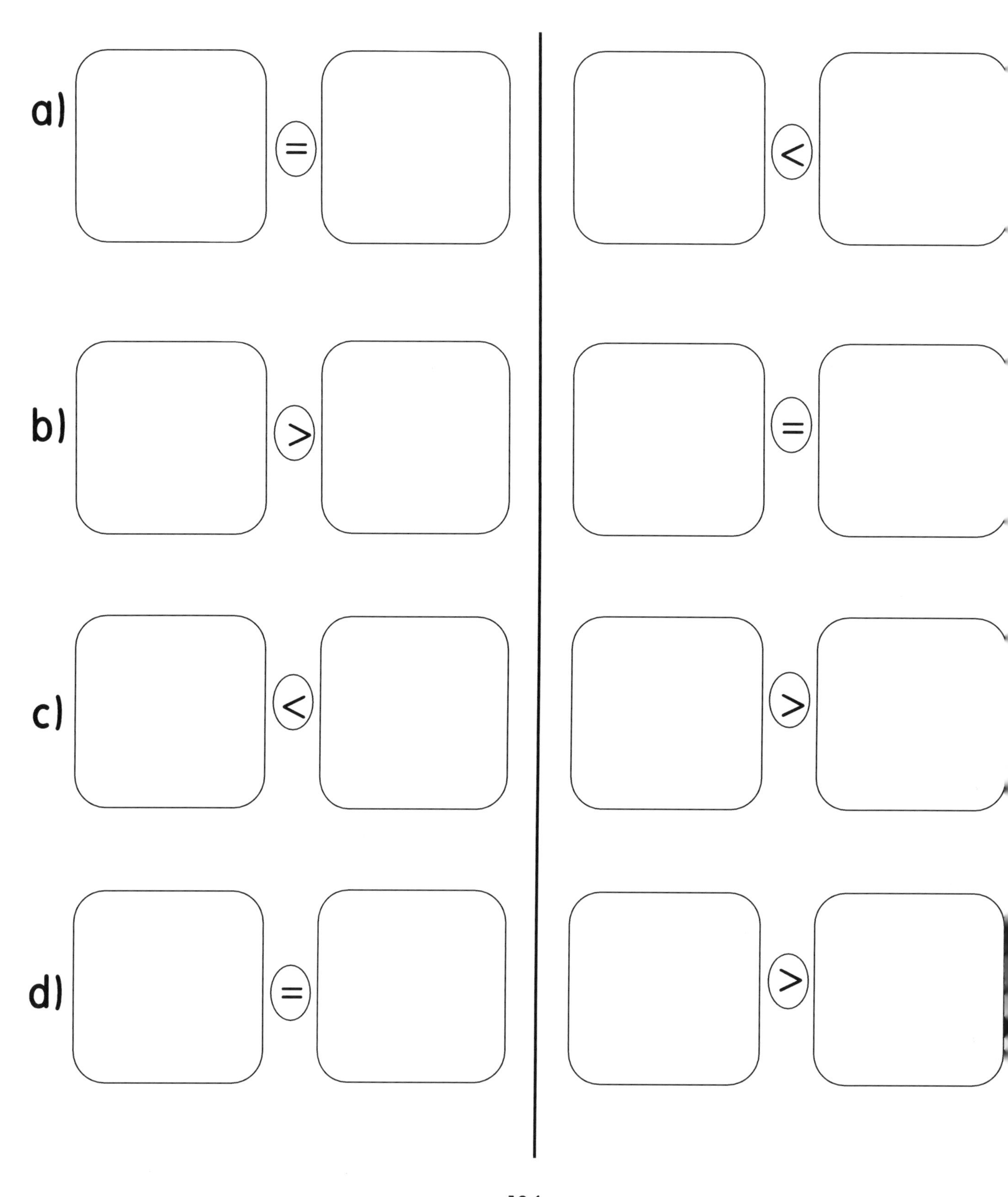

✵Read the numbers.

0 1 2 3 4 5

✵Now, write the number which comes just after.

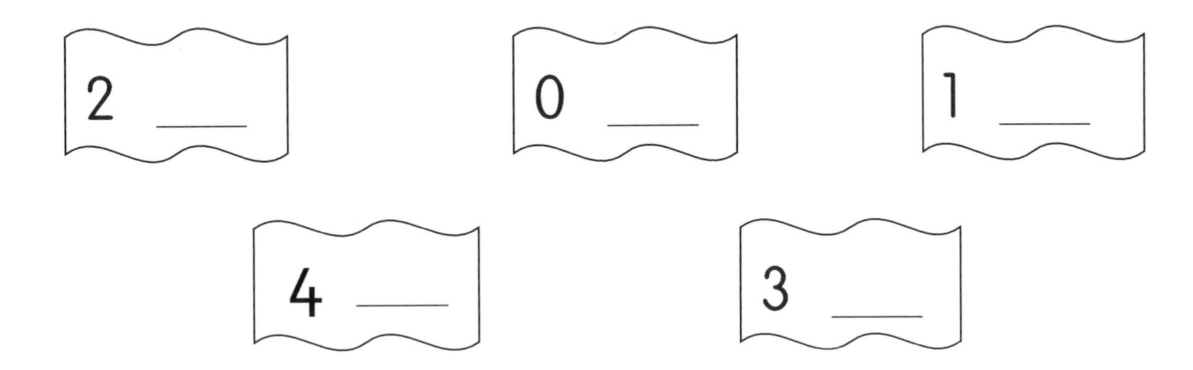

✵Write the missing numbers in the spaces.

a)

b)
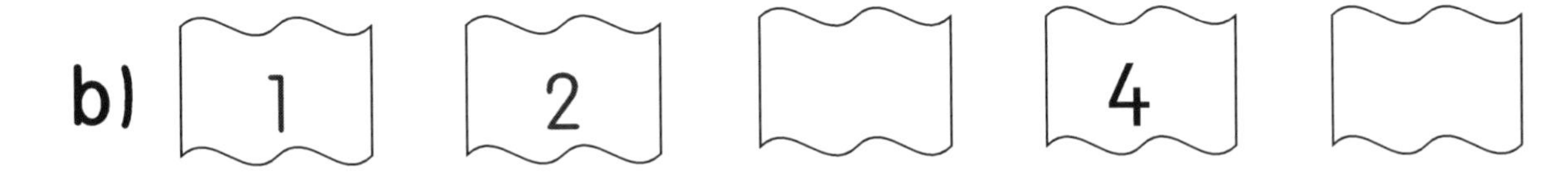

c)

✱Read the number names.

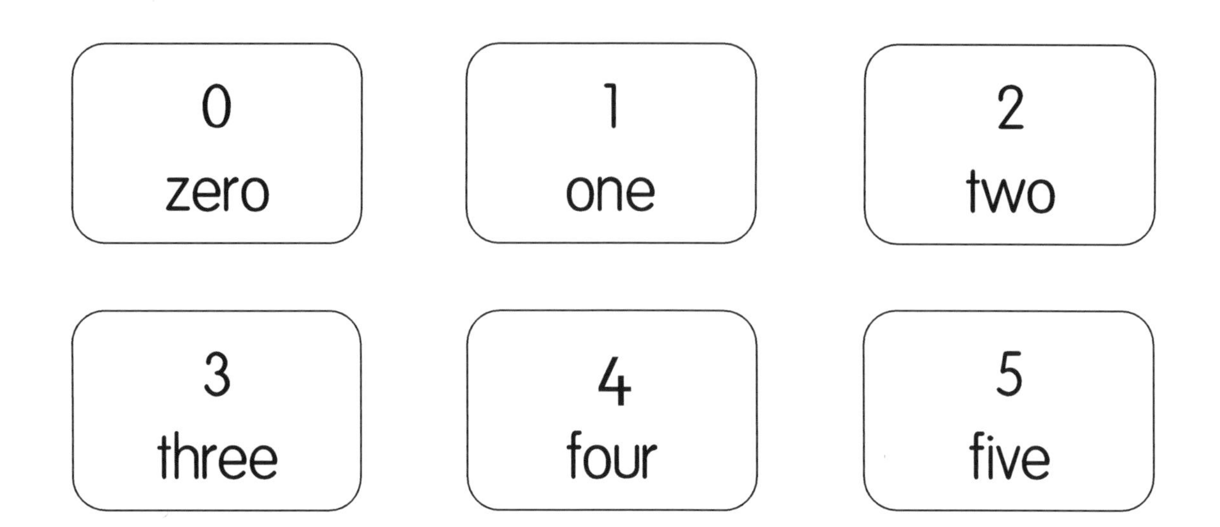

✱Now, match the number name to the number.

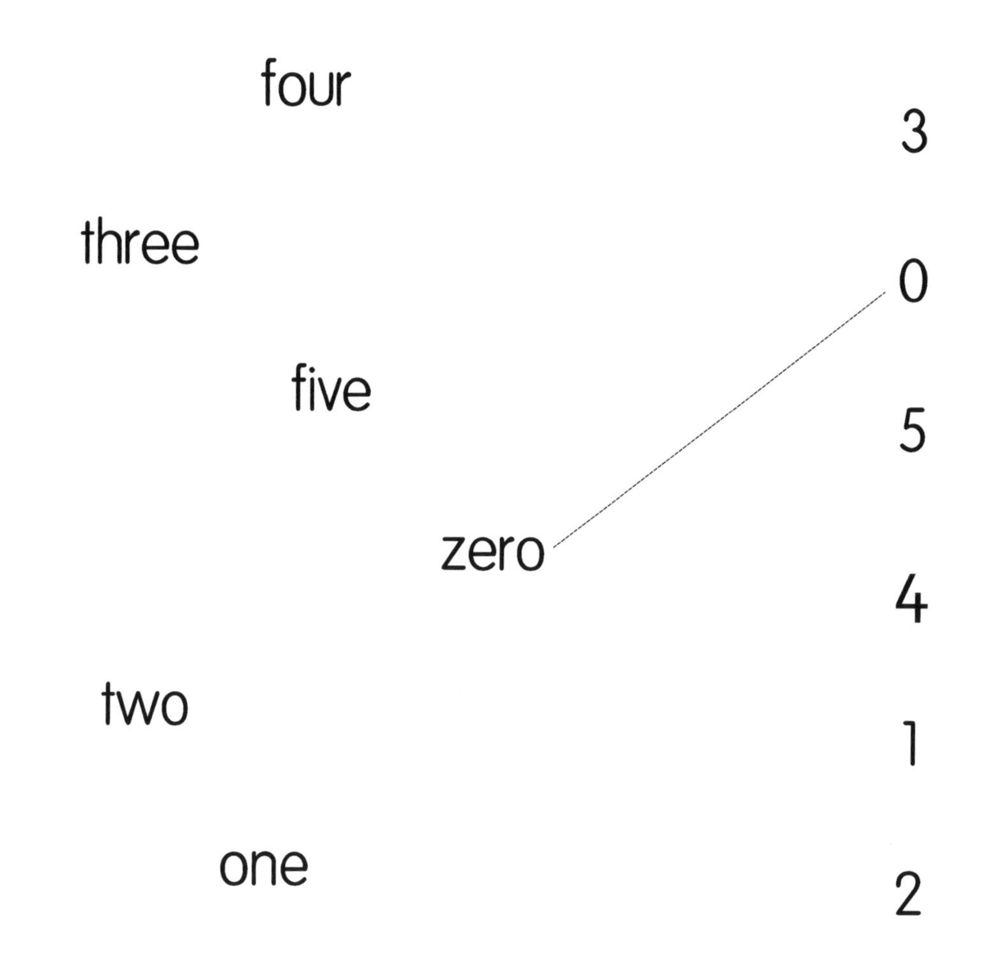

✱Write the missing letters for the number names.

0	1	2
z__r__	__n__	tw__

3	4	5
thr__ __	f__ __r	f__v__

✱Write the word for the number.

1 ______________ 4 ______________ 2 ______________

5 ______________ 0 ______________ 3 ______________

✱Read the numbers.

0 1 2 3 4 5 6 7 8 9 10

✱Now, write the number which comes just before.

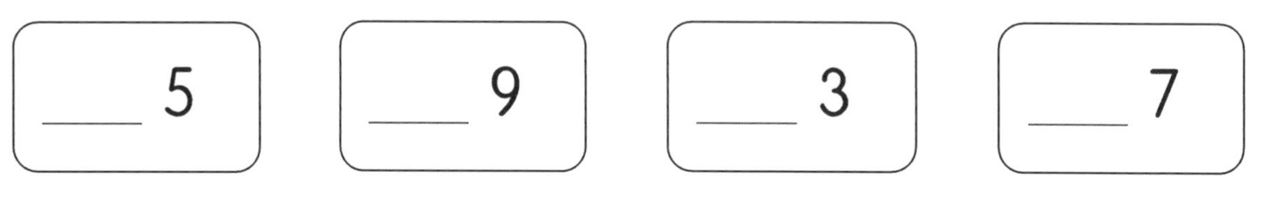

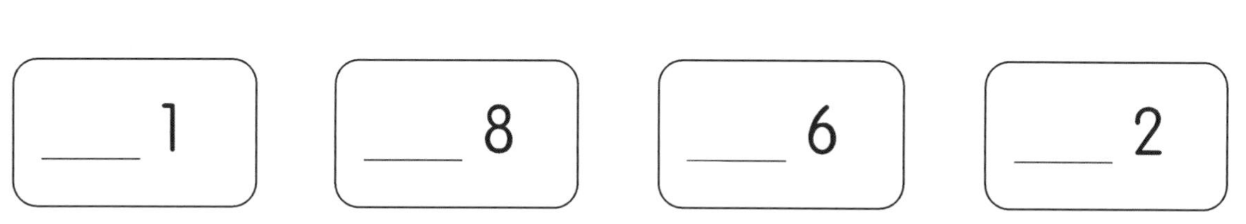

✱Write the missing numbers in the spaces.

a) 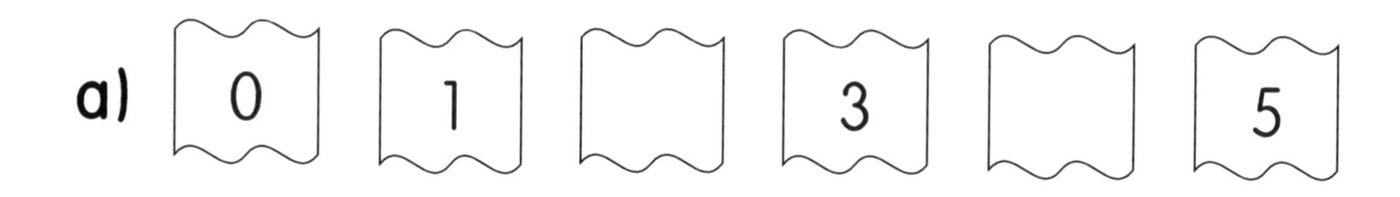

b) 3 4 6 8

c) 

✱Write the numbers from zero to ten on the line.

____

✱Now, write the number which goes between.

a) 

0 ____ 2

6 ____ 8

b) 3 ____ 5

7 ____ 9

2 ____ 4

c) 4 ____ 6

8 ____ 10

5 ____ 7

✵Read the number names.

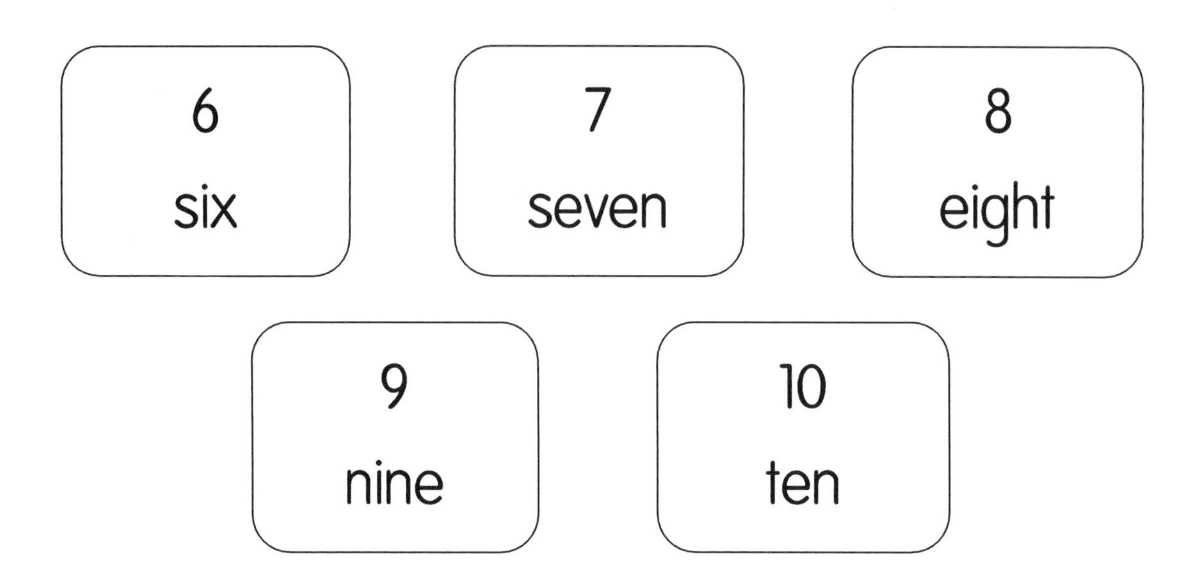

✵Now match the number name to the number.

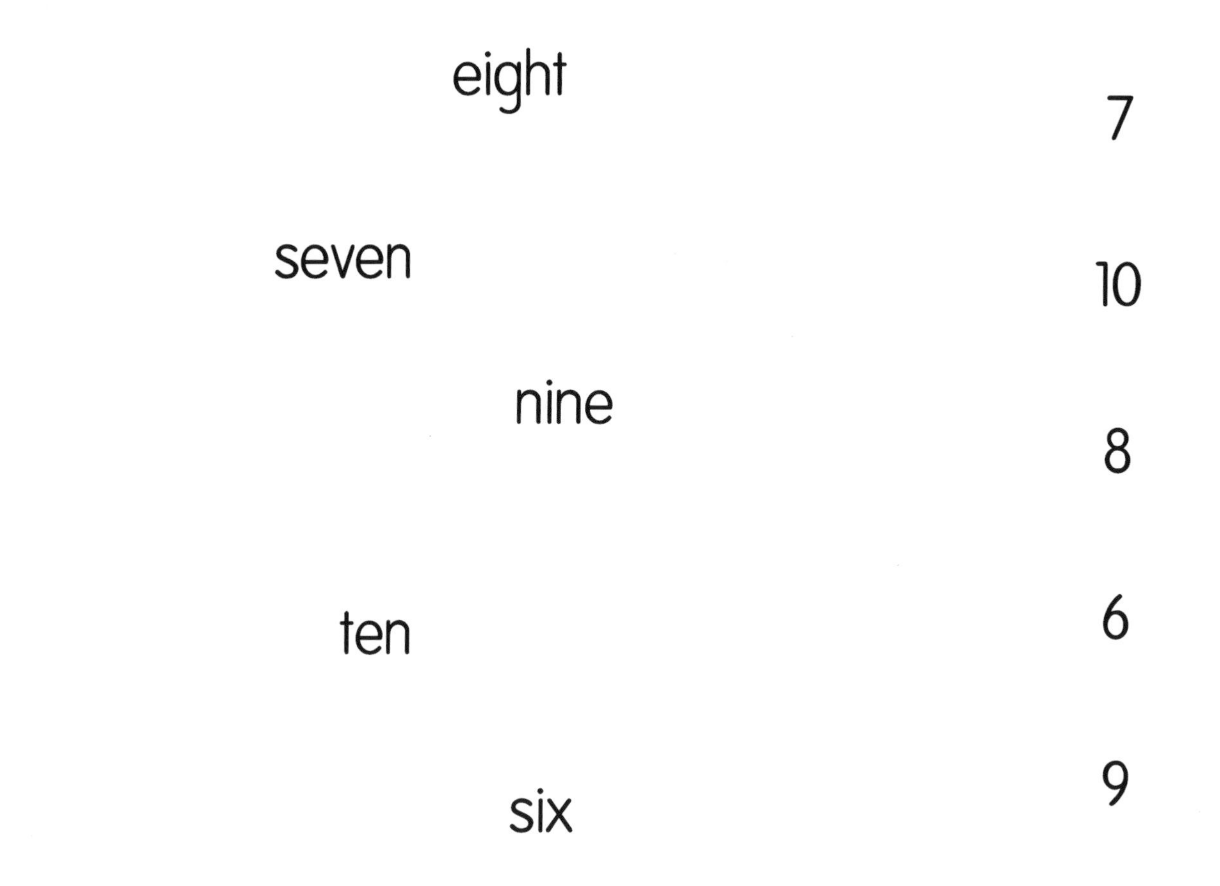

✵Write the missing letters for the number names.

10	6	8
t__n	s__x	e__g__t

7	9
s__v__n	n__n__

✵Write the word for the number.

7 ____________ 9 ____________ 6 ____________

8 ____________ 10 ____________

✱Read the position of the animals.

✱Now circle the position of the animals.

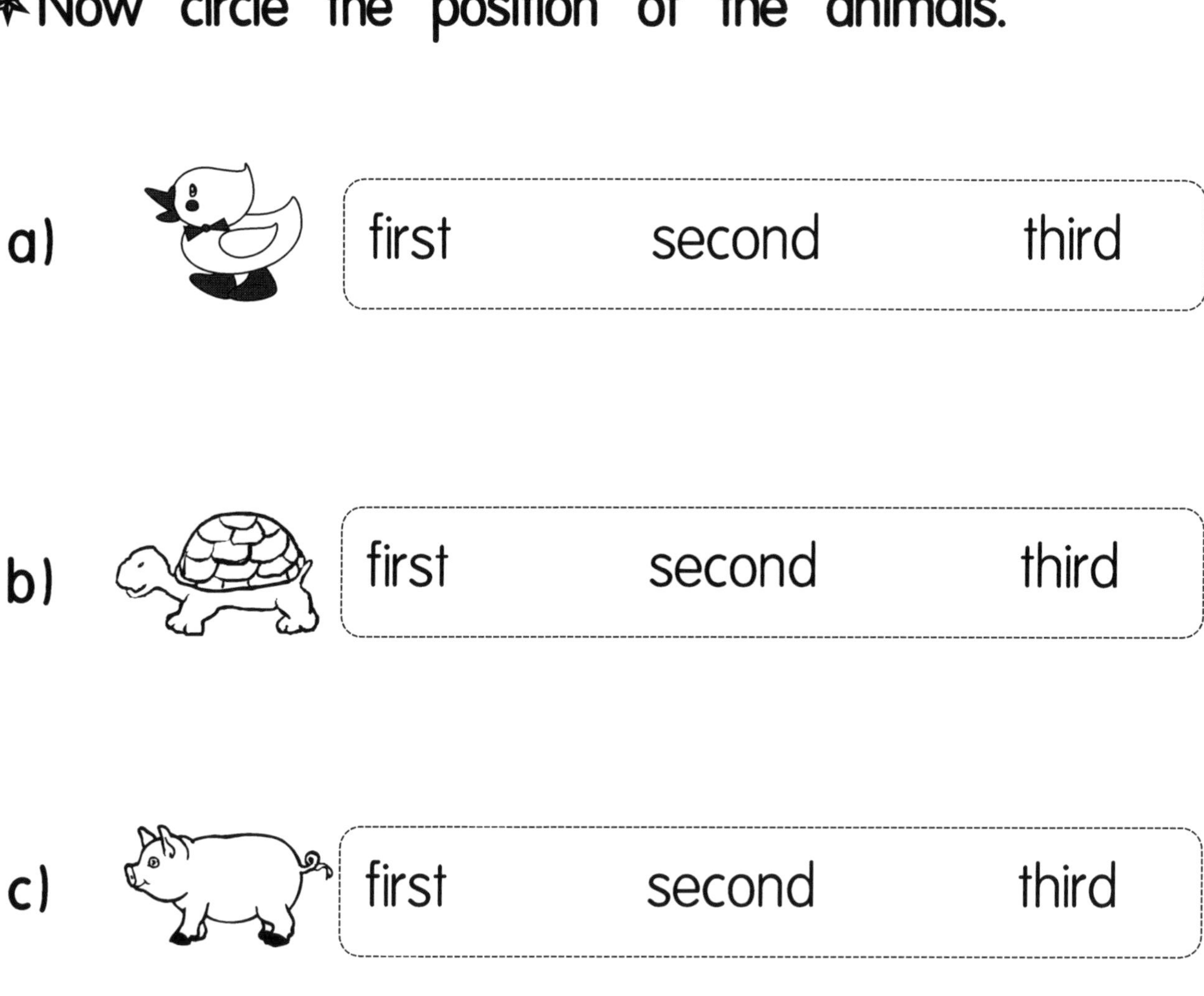

✷Write the word for the position.

1st ______________ 2nd ______________ 3rd ______________

a) Circle the second kite.

1st

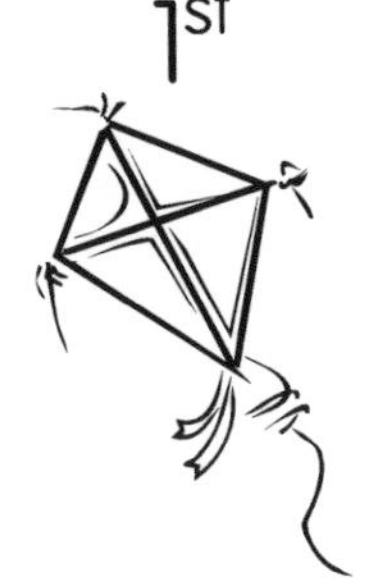 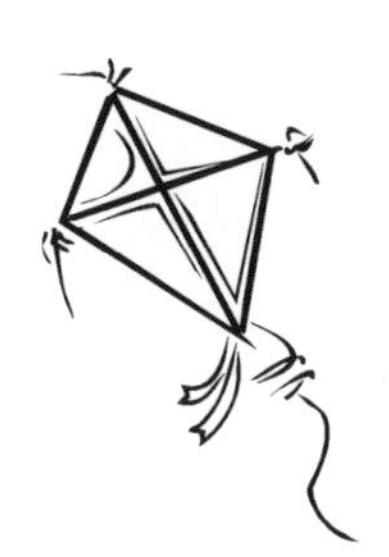 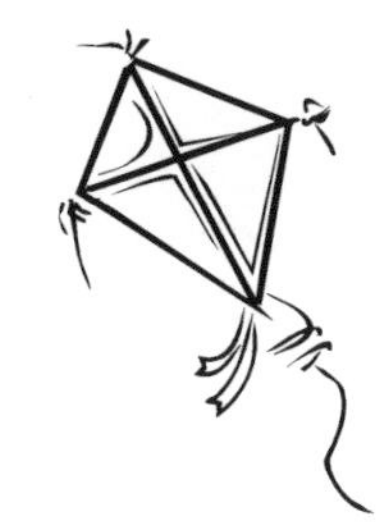 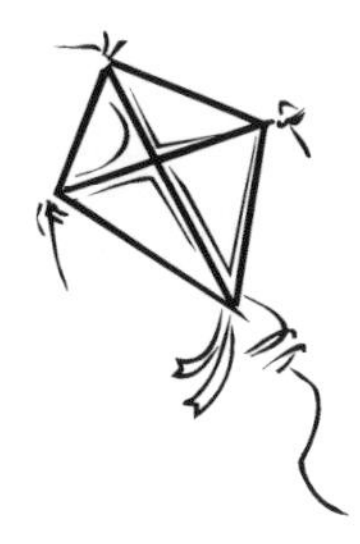

b) Circle the third boat.

2nd

c) Circle the first ball.

3rd

✷Write the position of the vehicles on the lines.

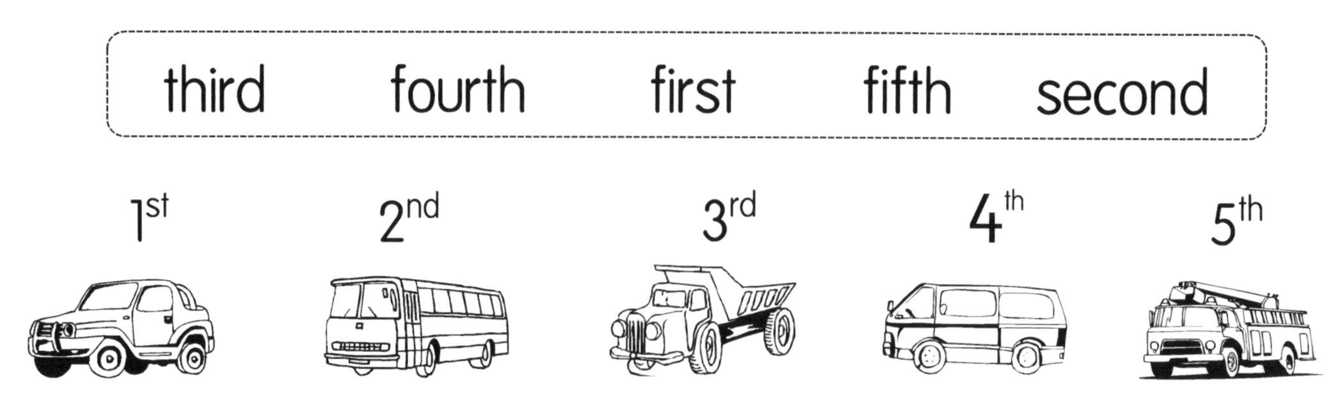

__________ __________ __________ __________ __________

✷Write the correct position in the space.

1) The doll is ________________.

2) The yo-yo is ________________.

3) The bear is ________________.

4) The kite is ________________.

a) Circle the second (2nd) fish.

b) Circle the fifth (5th) duck.

c) Circle the third (3rd) dog.

d) Circle the first (1st) horse.

e) Circle the fourth (4th) cow.

✷Circle the objects that are whole.

Put an 'X' on the objects that are fractions or parts.

✱Colour half of each shape.

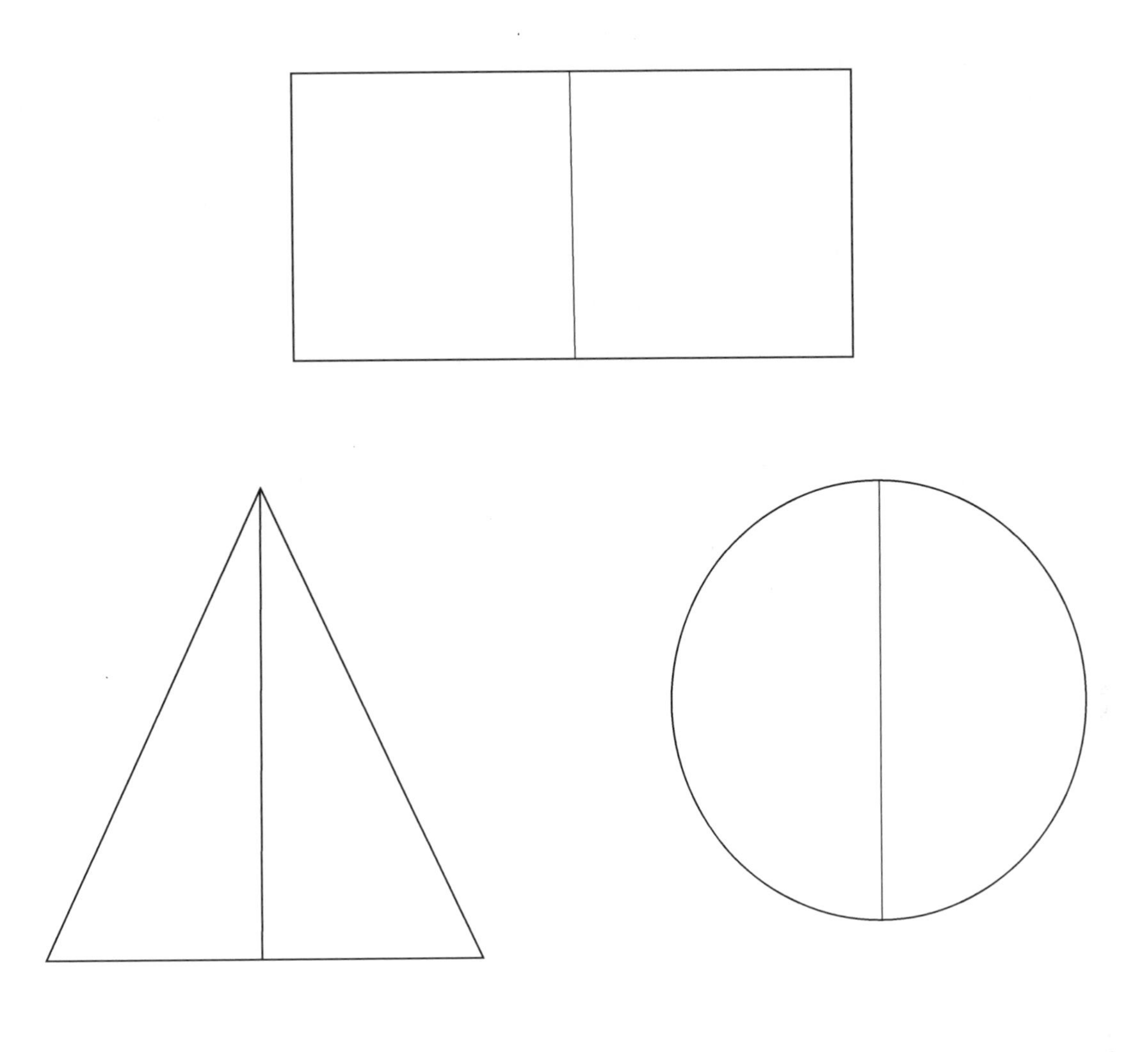

✱ Look at the pictograph and answer the questions.

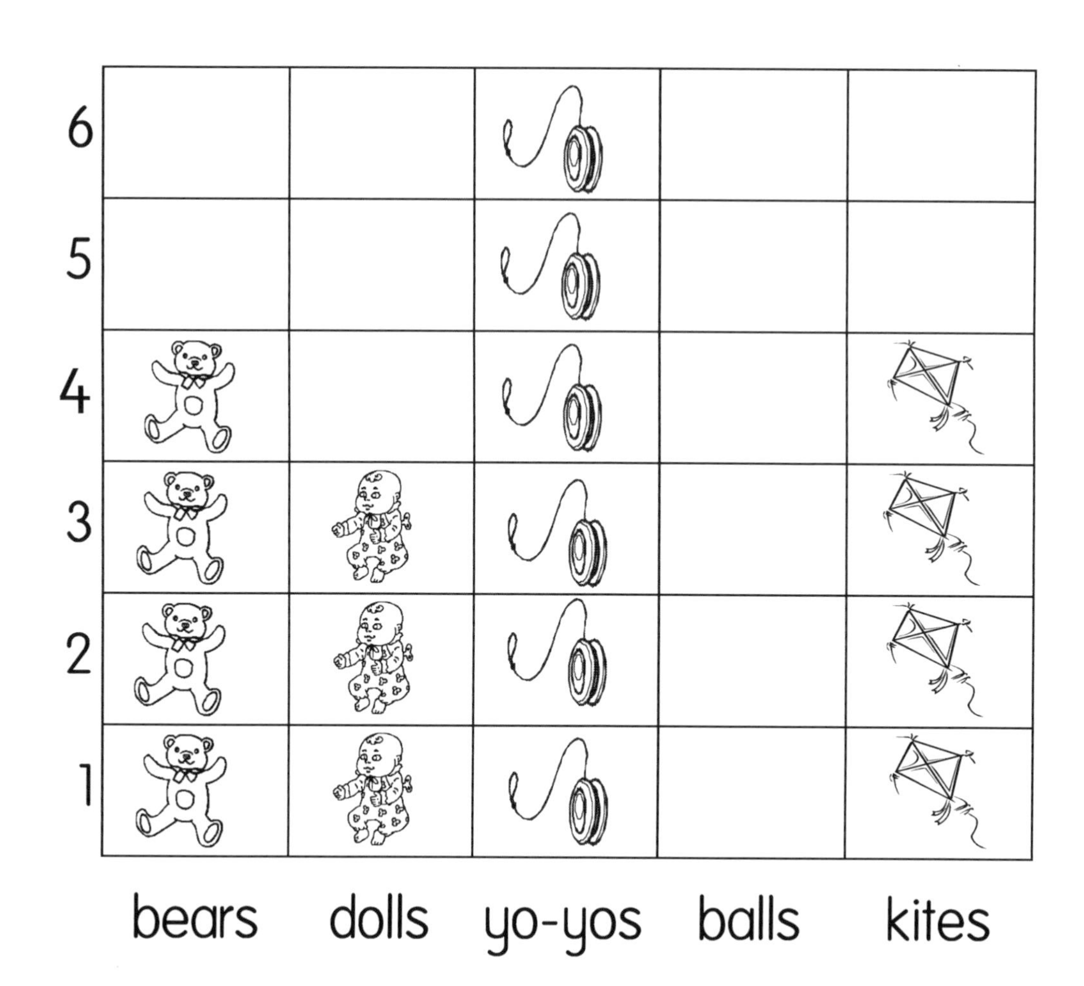

✱ Write how many.

✷ **Look at the pictograph and complete the sentences.**

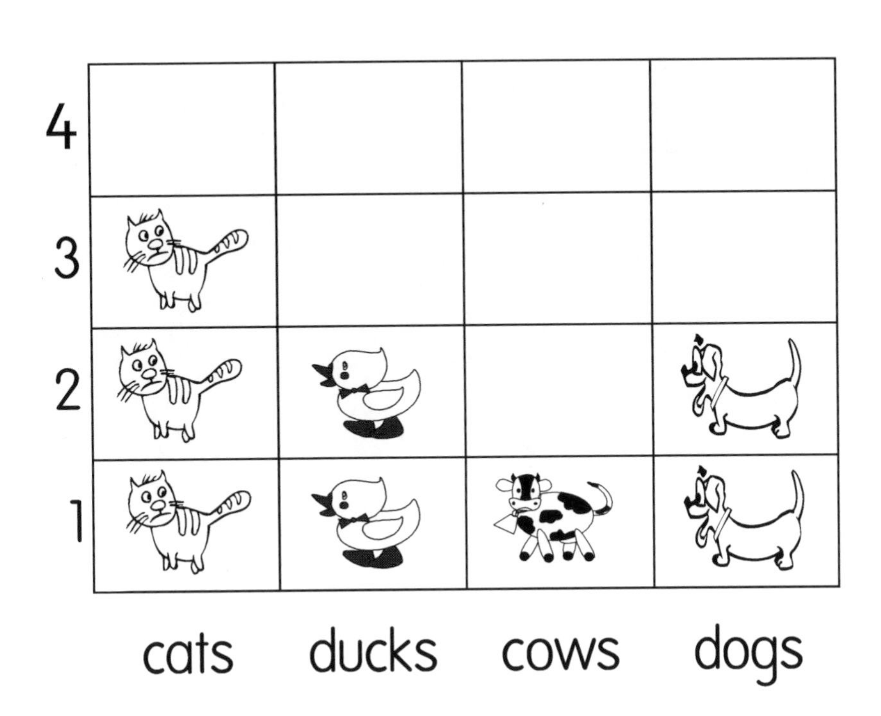

1. There is __________ cow.
2. The __________ are the most.
3. The __________ and __________ are the same.
4. There are __________ animals in all.
5. I like the __________.

✵Count and then write the correct amount on the line.

a)

b)

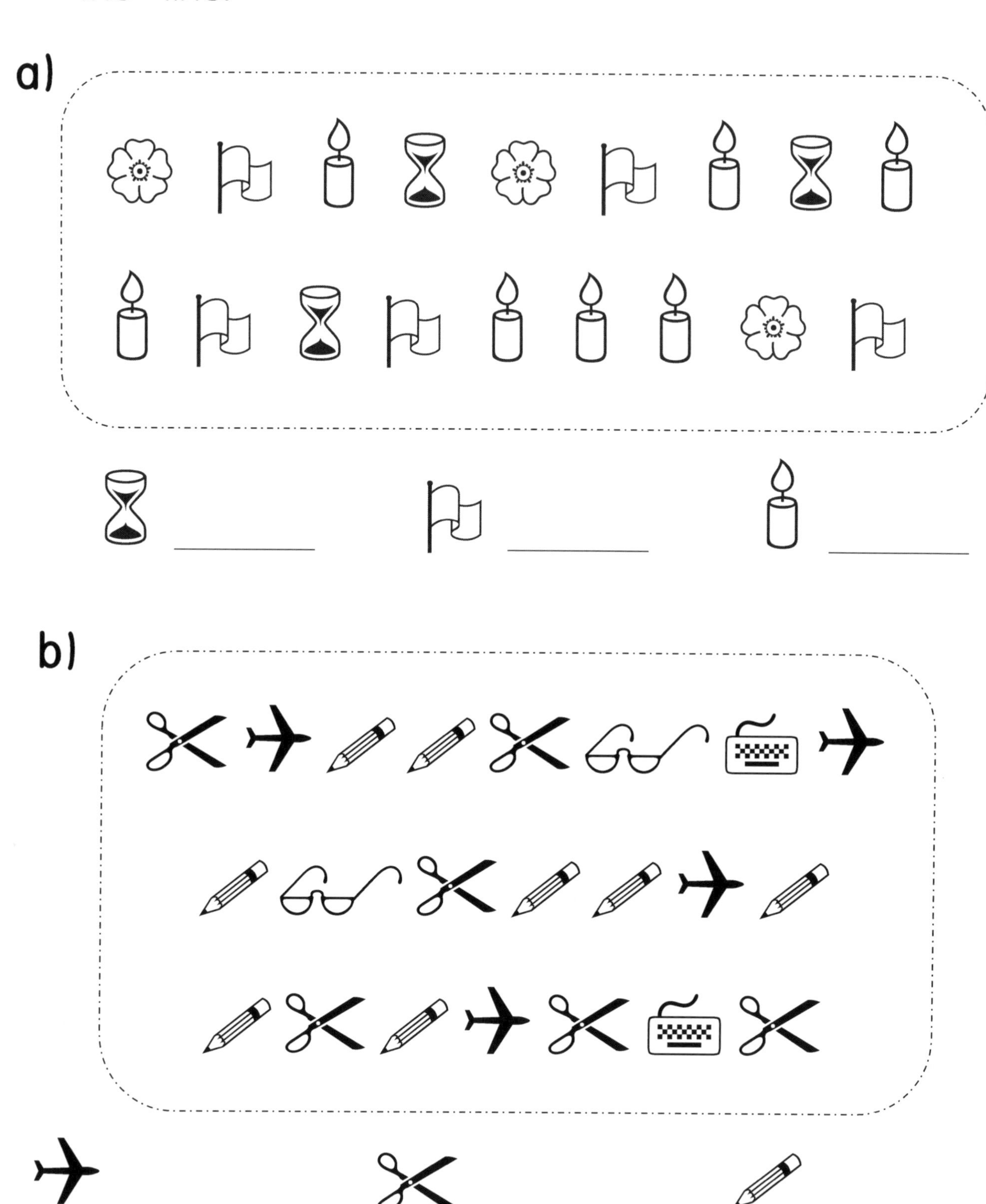

✱Look at the pictograph and put the correct sign in the boxes. (<   > =)

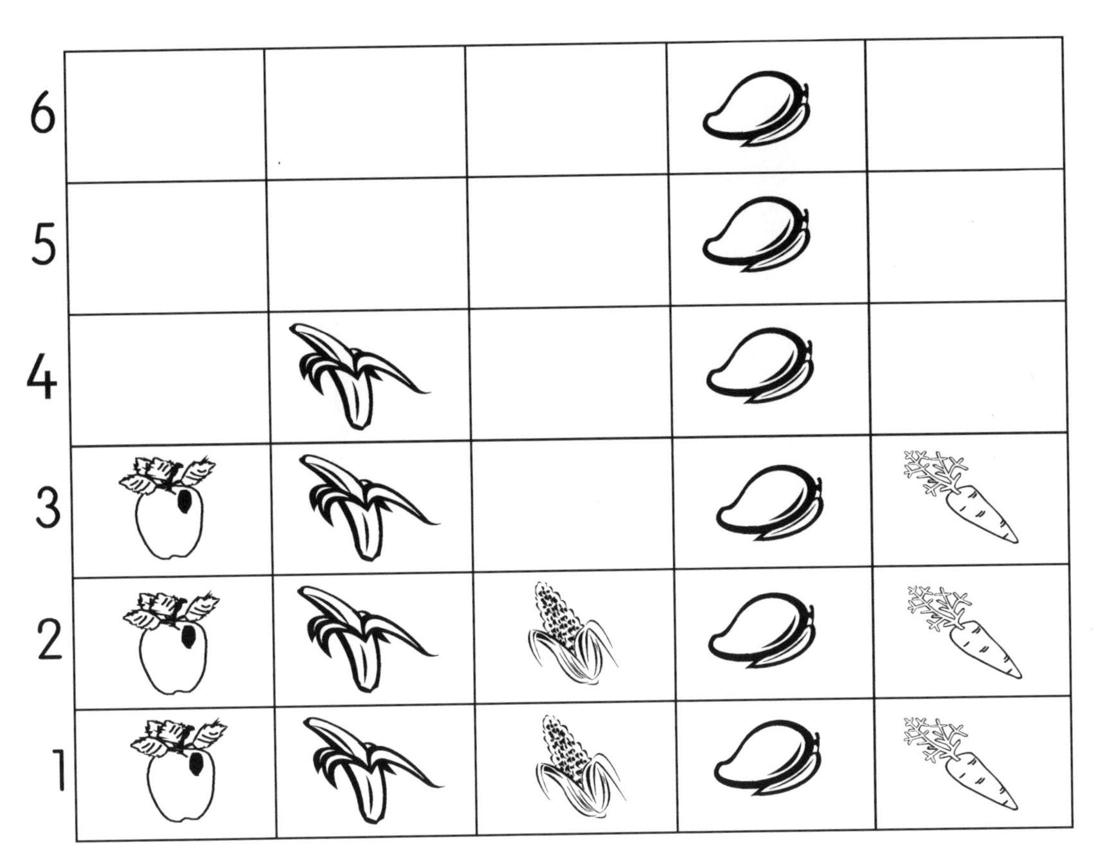

apples bananas corn mangoes carrots

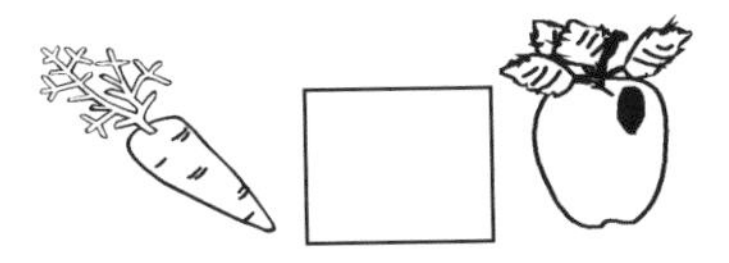

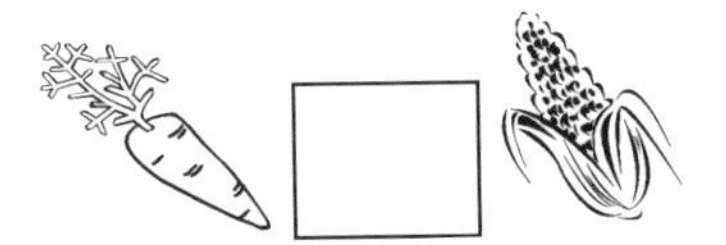

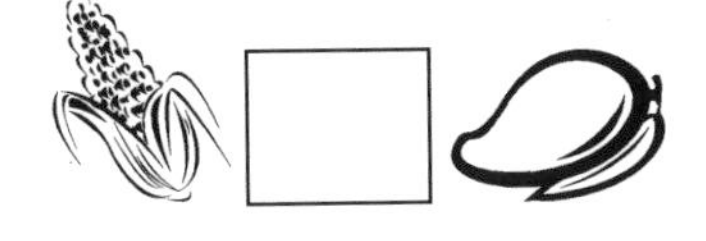

✱ Count the objects in the sets together and write the total in each box.

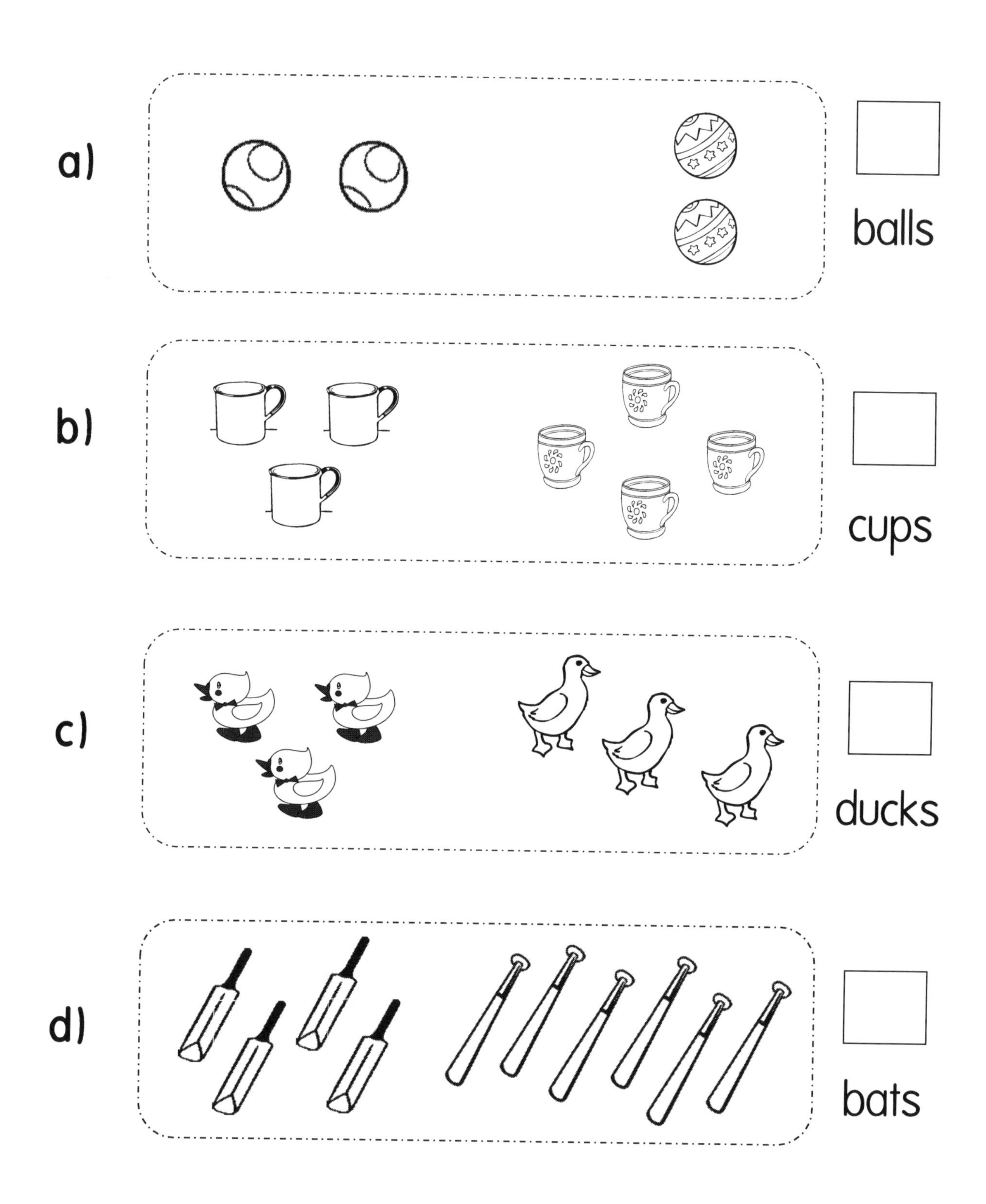

✱ Count the objects in the sets together and write the total in each box.

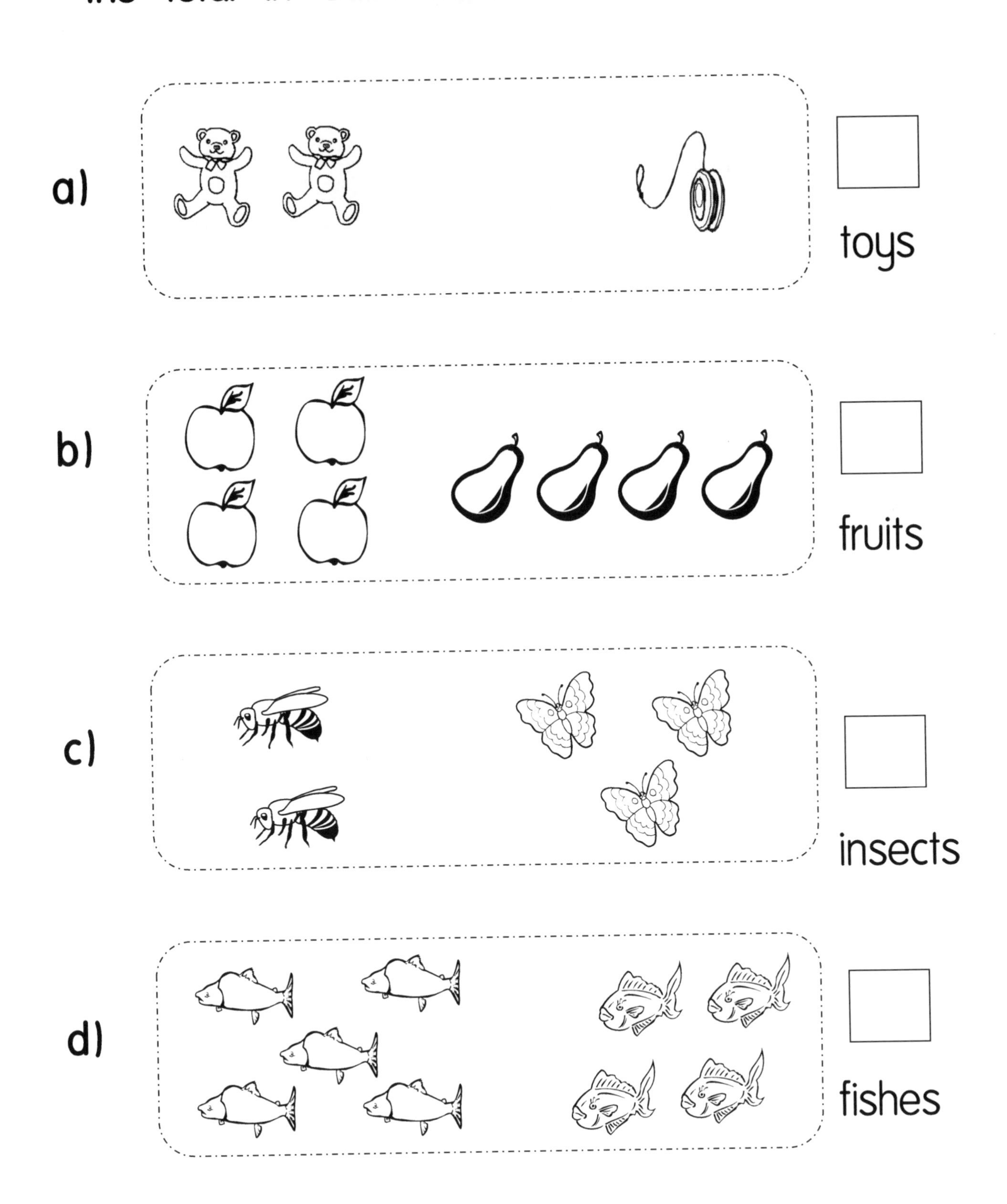

✱ Cross out the correct amount. Then write how many are left in each box.

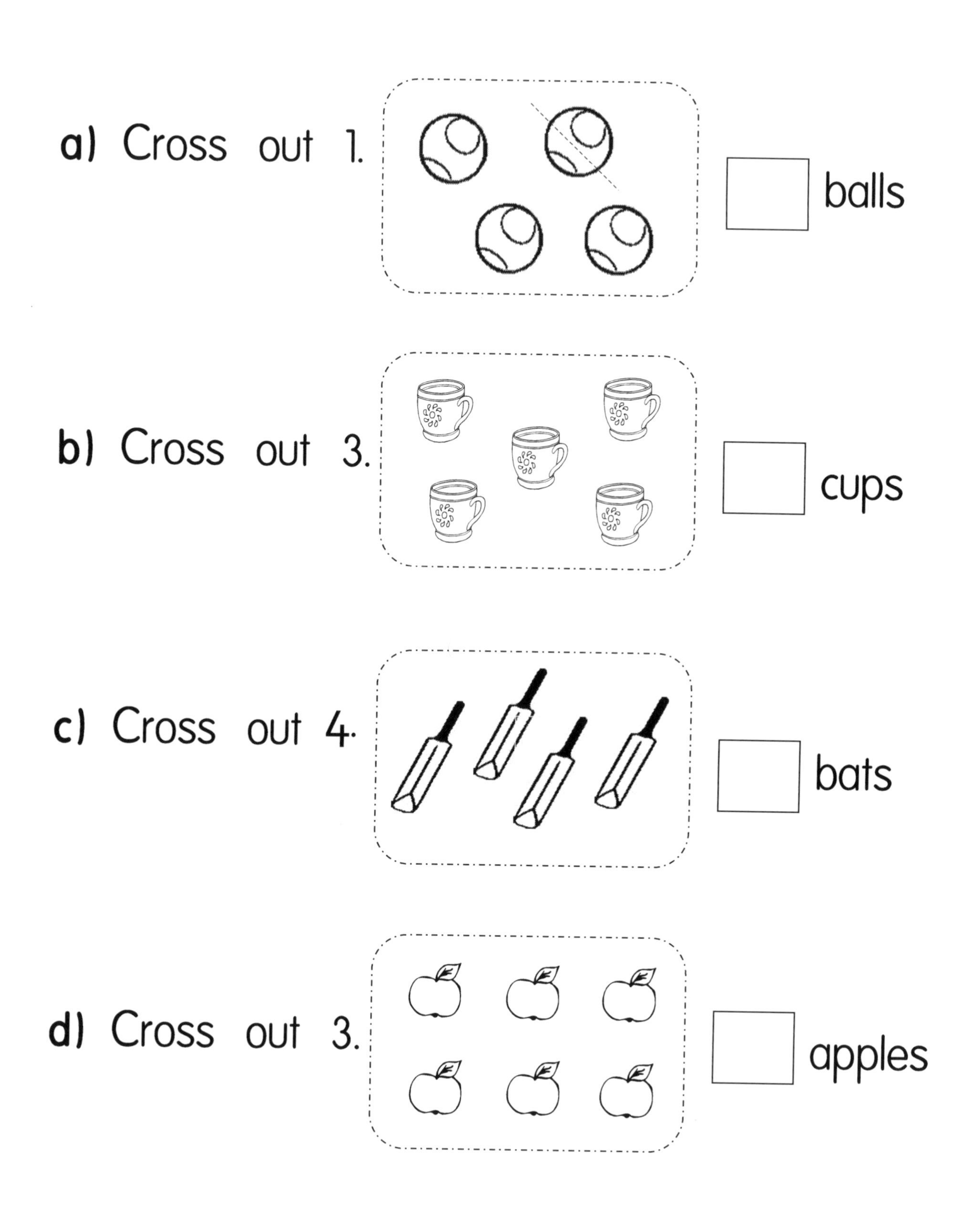

✱ Cross out the correct amount. Then write how many are left in each box.

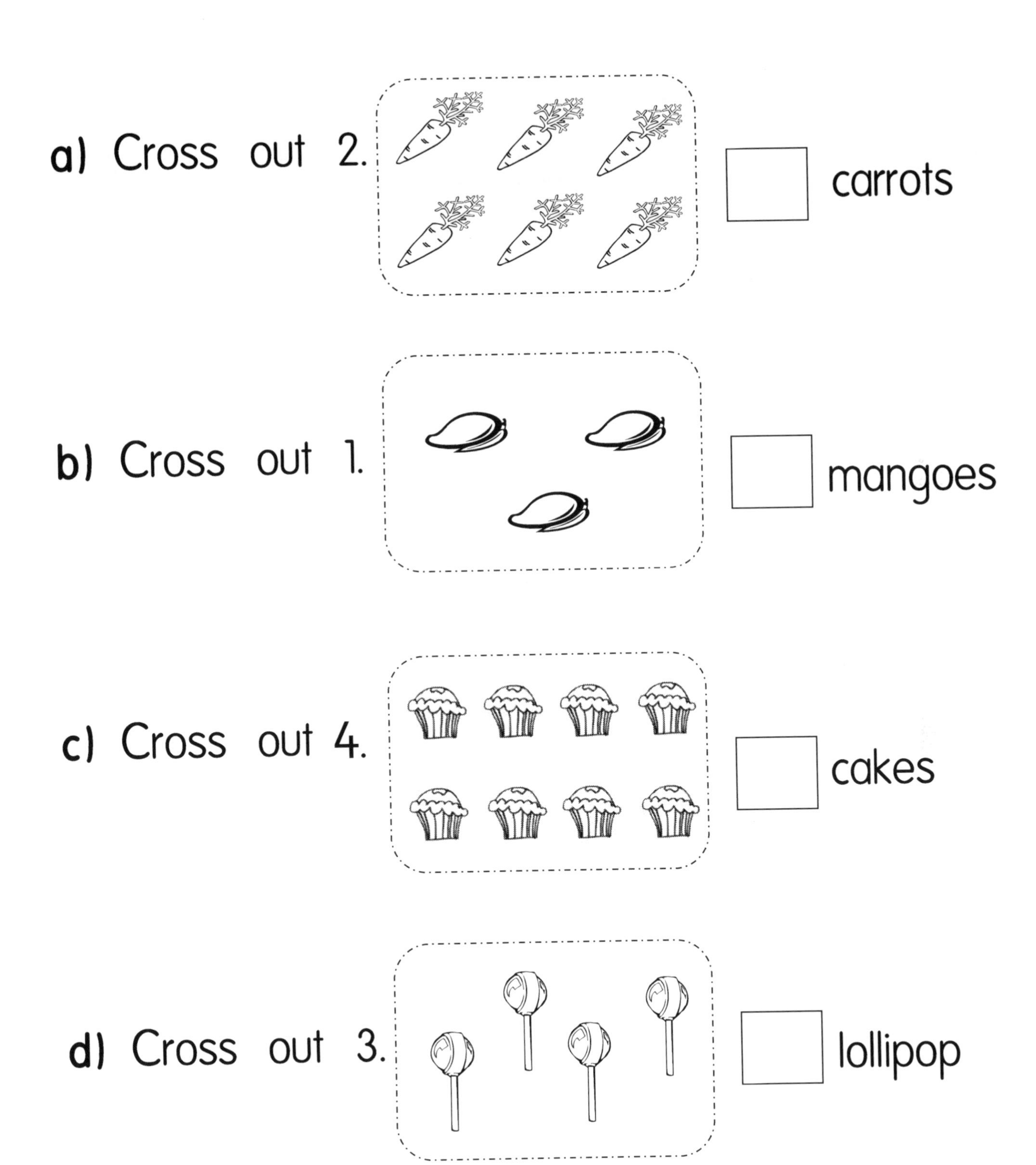

Homework

	Page / Pages
1	
2	
3	
4	
5	
6	
7	
8	
9	
10	
11	
12	
13	
14	
15	
16	
17	
18	
19	
20	

	Page / Pages
21	
22	
23	
24	
25	
26	
27	
28	
29	
30	
31	
32	
33	
34	
35	
36	
37	
38	
39	
40	

Made in the USA
Columbia, SC
07 September 2022